101 899 479 3
KT-221-082

The JCT Design and Build Contract 2005

The JCT Design and Build Contract 2005

David Chappell

Third Edition

Blackwell Publishing editorial offices:
Blackwell Publishing Ltd, 9600 Garsington Road, Oxford OX4 2DQ, UK
Tel: +44 (0)1865 776868
Blackwell Publishing Inc., 350 Main Street, Malden, MA 02148-5020, USA
Tel: +1 781 388 8250
Blackwell Publishing Asia Pty Ltd, 550 Swanston Street, Carlton, Victoria 3053, Australia
Tel: +61 (0)3 8359 1011

First published 1993
Second edition published 1999
Third edition published 2007 by Blackwell Publishing Ltd

2 2008

ISBN-13: 978-1-4051-5924-1

Library of Congress Cataloging-in-Publication Data

Chappell, David.
The JCT design and build contract 2005 / David Chappell. – 3rd ed.
p. cm.
Includes bibliographical references and index.
ISBN-13: 978-1-4051-5924-1 (hardback : alk. paper)
1. Construction contracts–England. I. Title.

KD1641.C485 2007
343.42′07869–dc22

2007006955

A catalogue record for this title is available from the British Library

Set in 9.5/11.5pt Palatino
by SNP Best-set Typesetter Ltd., Hong Kong
Printed and bound in Singapore
by Fabulous Printers Pte Ltd

The publisher's policy is to use permanent paper from mills that operate a sustainable forestry policy, and which has been manufactured from pulp processed using acid-free and elementary chlorine-free practices. Furthermore, the publisher ensures that the text paper and cover board used have met acceptable environmental accreditation standards.

For further information on Blackwell Publishing, visit our website:
www.blackwellpublishing.com

Contents

Preface to the Third Edition

The JCT Standard Form of Building Contract With Contractor's Design has proved very popular. Although not the only design and build contract, it was certainly the best known. Design and build is still a very popular form of procurement and there is little sign that the trend is decreasing. The popularity of design and build owes a lot to the perception of a single point responsibility, virtually guaranteed price and time and reduced claims opportunities if certain key principles are observed. The extent to which that perception is justified will become clear.

Originally produced in 1981, the contract form became known as CD 81. It was reprinted incorporating substantial amendments as WCD 98 in 1998.

In almost eight years since the second edition of this book, many things have changed. The main change is that, in 2005, the Joint Contracts Tribunal completely revised the contract. It is now straightforwardly called the Design and Build Contract (DB). The structure of the contract is dramatically changed, with the introduction of contract particulars at the front of the contract to replace the several appendices in the previous edition; schedules have been added at the end of the contract; and third party rights and warranties have been introduced. The clauses have been restructured, re-numbered and re-worded, sectional completion has been incorporated into the wording of the contract and some of the terminology has been changed. This edition of the book takes account of all these changes. The RIBA Terms of Engagement SFA/99, CE/99 and the Amendments DB1/99 and DB2/99 for use with a design and build contract have also been replaced with entirely new terms of engagement and the effect of these new terms is considered. At the time of writing, they are still in draft. The opportunity has also been taken to review the standard novation agreements which have been produced by the Construction Industry Council and by the City of London Law Society. The effect of the CDM Regulations 2007 and the JCT Amendment 1, dealing with the Regulations and changes to third party rights and warranty provisions, has been taken into account. The amount of case law has also increased. Over 70 cases have been added to the text. The opportunity has been taken, where appropriate, to reconsider earlier conclusions in the light of the new contract.

The structure of DB is very similar to the new Standard Building Contract (SBC), as WCD 98 was to JCT 98. Indeed, much of the wording is identical. This may again lead to problems as the parties may overlook the very many subtle, and some quite clear, differences. Misunderstandings and disputes may result. Architects acting as employer's agents often try to deal with the contract as though it were the traditional standard form. That is a recipe, if not for disaster, at least for substantial claims. The fact is that DB, as WCD 98 before it, is a very complex document. Problem areas are still:

- The allocation of design responsibility
- Discrepancies

- The role of the employer's agent
- Payment provisions
- The approval of drawings.

This book is designed to operate on two levels: as a practical guide to assist the user in what to do next, and as an authoritative text with references to appropriate case law. Where the meaning of the contract is obscure and judicial pronouncements offer no guidance, a view has been taken and advice given appropriate to the situation. The text is illustrated, where possible, with examples of the way the contract works in use.

A common method of writing about building contracts is to provide a commentary clause by clause. Although that approach has the advantage of concentrating on individual clauses, it seems quite sterile in its effect when essentially the contract must be read as a whole. It has long seemed sensible and of far more use to the practitioner to deal with the contract on a topic basis, referring to relevant clauses as appropriate. Therefore, this book proceeds by examining the roles of the participants and then considering particular important topics such as termination, claims and payment. For ease of reference, some of the information is also provided in tabular form.

It is hoped that the book will be useful for employers about to embark on design and build for the first time, as well as to the contractor, to the professional acting as employer's agent, whether architect, engineer or surveyor, and to the design team acting for the contractor.

The first edition of this book was written with the late Professor Vincent Powell-Smith, an authority on construction contracts. Although his name was retained on the cover of the second edition, that was simply to acknowledge that much of the text still bore his hallmark. My thanks to Michael Dunn BSc(Hons), LLB, LLM, FRICS, FCIArb, who has provided assistance in various practical ways, and to Michael Cowlin LLB(Hons), Dip Arb, Dip OSH, FCIArb, Barrister, who gave valuable advice in relation to the Final Account and Final Statement.

Throughout the text the contractor has been referred to as 'it' on the basis that it is a corporate body.

Although the clauses, in what the JCT insists on calling the 'Conditions', are referred to as 'Sections', throughout this text they have been referred to as 'clauses'. Thus clause 6.3 is part of clause 6, 8.2 is part of clause 8 and so on. Not only does this seem clearer than referring to clause 6.3 as part of section 6 etc., it also avoids confusion between sections of the 'Conditions' and sections of the chapters of the book.

David Chappell
Wakefield
April 2007

Chapter 1
Introduction

1.1 Definitions

In the traditional procurement scenario, an employer appoints an architect to design a building. The architect prepares designs, seeks approval from the client and steers the project through all the stages of what is commonly known as the RIBA Plan of Work (currently and confusingly again under revision). This includes obtaining planning permission, seeking tenders, dealing with the contract and administering the contract during operations on site. Throughout, the architect acts for the client and gives the client a professional service, perhaps modified to suit particular client preferences. Essentially, design is in the hands of the architect who develops it into production information, while construction is carried out by the contractor precisely in accordance with the architect's designs. This is still the single most popular category of procurement of buildings in the UK, although within the category there are variants such as management contracting, construction management, project management, etc.

Design is a difficult concept to define. It has any number of connotations, as the various dictionary definitions make plain. It can be 'a preliminary plan or sketch for the making or production of a building' as well as 'the art of producing these'. Design may be a scheme or plan of action and it can be applied equally to the work of an architect in formulating the function, construction and appearance of a building as to an engineer determining the sizes of structural members, and clearly it involves the selection of materials suitable for the purpose of the proposed structure. It is generally accepted that an architect is designing, not only when producing presentation drawings showing the way the building will look, but also when producing constructional or working drawings showing how the component parts of the building fit together. The architect is also most certainly designing when producing large-scale details of various parts of the building and when preparing the detailed written specification. In *Rotherham MBC* v. *Frank Haslam Milan & Co* (1996) CA, the architect specified the use of a fill material which subsequently gave rise to problems. The Court of Appeal held that the extent to which the contractor was free to choose was not to enable the contractor to exercise skill and judgment, but because the architect believed that further stipulations were unnecessary.

On the other hand, the contractor is not designing when it puts the components together in a way and using materials specified by the architect. Yet the contractor may be involved in some design even in a building erected under a traditional procurement system. Consider a piece of built-in joinery designed by the architect as part of the building. It may be designed in great detail with full-size sections through its parts, but it is still likely that the joinery will not be designed in every detail. If there is any portion not so designed, it is possible that the contractor will assume some design responsibility if it carries on and produces what it assumes will be required rather than asking the architect for more information: *C G A Brown Ltd* v. *Carr and Another* (2006).

In the Australian case of *Cable (1956) Ltd* v. *Hutcherson Bros Pty Ltd* (1969), for example, although the contractor had tendered for the design, supply and installation of a bulk storage and handling plant to be built on reclaimed harbour land, the contract required the contractor's drawings to be approved by the employer's engineer. The drawings as approved showed ring foundations for storage bins. When these were erected and filled, subsidence occurred. The High Court of Australia held that, on the true interpretation of the contract documents, the contractor was not liable as it was not responsible for the suitability of the design. In the court's view, the contractor 'promised no more than to carry out the specified work in a workmanlike manner' and it would appear that the employer had not in fact relied on the contractor's skill and judgment in respect of the design.

In *Brunswick Construction* v. *Nowlan* (1974), however, Nowlan engaged an architect to design a house and then contracted with Brunswick to erect it to the architect's design. No architect supervised the construction. The design was defective and made insufficient provision for ventilating the roof space. The Supreme Court of Canada held the contractor liable for a resultant attack of dry rot, on the basis that an experienced contractor 'should have recognised the defects in the plans . . . knowing the reliance which was being placed upon it'. It should have been obvious to the builder that the building would not be reasonably fit for its intended purpose if it was constructed in accordance with the defective plans.

Even if the architect remembers to draw sections through every portion, it is very unlikely that details of the screws holding everything together will be included. The architect will assume, probably correctly, that the joiner will know the kind of fixings, sizes, materials and spacing required. This is commonly referred to as 'second order design'. Architects vary in the amount of second order design they leave to the contractor and it is very difficult in some instances to decide what is the difference between second order design and workmanship. In practice, this can lead to problems in allocating responsibility where traditional procurement paths are taken. Two cases have considered the meaning of 'faulty design' and 'faulty workmanship', albeit in the context of insurance. The Australian High Court, in *Queensland Government Railways and Electric Power Transmission Pty Ltd* v. *Manufacturers' Mutual Insurance Ltd* (1969), set out the difference as follows:

> 'Faulty workmanship I take to be a reference to the manner in which something was done, to fault on the part of a workman or workmen. A faulty design, on the other hand, is a reference to a thing. If the words were "faulty designing" the two phrases might perhaps be comparable: but the words are "faulty design". I think that, reading those words in their ordinary meaning, the collapse of the piers was the result of their design being faulty.'

This judgment was noted with approval in a subsequent English case (*Kier Construction Ltd* v. *Royal Insurance Co (UK) Ltd and Others* (No.1) (1992)), where the judge said:

> 'When one talks of defective workmanship one is condemning the workman, but when one speaks of an object being defective in workmanship . . . , one may not necessarily make the same condemnation. A skilled and careful workman may produce an object which can be said from an objective point of view to be defective in workmanship, as a result of carefully following a detailed statement as to his method of work,'

and later:

> 'Workmanship is the skill required to convert a design plan and specification into an object . . . The first duty of good workmanship is to follow the design plan and specification . . . The specification itself is not workmanship, it is a requirement which workmanship has to follow.'

Although these comments are useful in differentiating between design and workmanship, it appears that the courts have not quite succeeded in identifying the precise point at which design stops and workmanship begins. Where the contractor is responsible for both design and workmanship, as under DB, the problem is only likely to arise where a considerable amount of design is carried out by the employer's design team before tender.

The idea of design and build is that the design and the construction of a project are in the hands of one firm. This appears to make eminent sense in that, in theory at any rate, it results in one point responsibility. In practice, it is not so simple. There are many terms which seem to be used indiscriminately for design and build. There are differences. The main types of design and build are as follows:

- *Design and build:* The contractor takes full responsibility for the whole of the design and construction process from initial briefing to completion of the project. This is the term which the industry tends to use as the general name for all variants of this procurement category.
- *Design and construct:* This is a wider term and it includes design and build, but also other types of construction such as purely engineering works of various kinds.
- *Develop and construct:* This is a term which lacks precision, but which is often used to describe a situation where a contractor is called upon to take a design which is partially completed and to develop it into a fully detailed design before being responsible for construction. Whether, in such a situation, the contractor is responsible for the original design as well as the development work will depend upon the precise terms of the contract. However, where this type of design and build is carried out on a simple exchange of letters, it is probable the contractor is responsible for the whole of the design.
- *Package deal:* Strictly, this term suggests that the contractor is responsible for providing everything. It particularly refers to systems of industrialised buildings which can be purchased and erected as a 'package'. The employer will usually be able to view similar completed buildings before proceeding. Closed systems of industrialised building are indicated.
- *Turnkey contract:* This is a procurement method in which the contractor really does do everything, including providing the furniture if required. The idea is that when the employer takes possession, all that remains to be done is to turn the key. It has been said that this is not a term with a precise legal meaning: *Cable*

(1956) Ltd v. *Hutcherson Bros Pty Ltd* (1969). However, another view is that the use of such term in contract documents is likely to indicate that the contractor is undertaking at least some design responsibility.
- *Design and manage:* This is not strictly design and build at all, but simply an architect-led version of the contractor-led construction management.

It is, however, important to determine whether the contract is a traditional one or a true 'design and build' contract because where the contractor offers not only to undertake the construction work but also to perform some or all of the design duties usually undertaken by the employer's professional team, then unless the express terms of the contract provide otherwise, the 'design and build contractor' will be under an obligation to ensure that the building as designed is suitable for its intended purpose.

This is well illustrated by *Viking Grain Storage Ltd* v. *T H White Installations Ltd* (1985) where contractors undertook to design and build a grain storage and drying installation. The installation was defective. The plaintiffs alleged that some of the materials used were defective, that some of the construction work was badly performed, and that aspects of the design were unsuitable. The installation was not fit for its intended purpose. On a preliminary issue, it was held that the defendants were strictly liable. The fact that they had used reasonable care and skill was no defence. There was an implied obligation that the finished installation would be fit for its intended purpose. Nothing in the express terms of the contract contradicted this obligation. The design and build contractor's liability is, in the absence of an express term to the contrary, equivalent to that of a supplier of goods, the only proviso being that the employer must have relied on its skill and judgment.

1.2 Advantages and disadvantages

The advantages of design and build are usually said to be as follows:

- The employer can refer to a single point of responsibility throughout the procurement process and after construction is complete if there are any latent problems. This is in contrast to the traditional systems where the employer's point of contact is the architect, but if there are difficulties, responsibility may lie with any one or more of a range of firms including the contractor, the architect, quantity surveyor, engineer and other consultants.
- There is less risk, because the cost is virtually guaranteed and there is a better than average chance of meeting a fixed completion date.
- The total procurement period is likely to be shorter than a similar project using traditional methods. This is because the contractor is in charge of the whole process.
- Except when DB is used, the contractor undertakes that the finished building will be fit for its purpose.
- The design concept is likely to be more easily buildable.
- There are likely to be fewer claims, because the factors which commonly trigger such claims are mainly under the control of the contractor.

Disadvantages are said to be as follows:

- The employer will have less control than under a traditional system.
- The system is not flexible. If the employer makes any changes in the requirements, it opens the door to claims for extensions of time and direct loss and/or expense.
- The Employer's Requirements must be prepared carefully so as to accurately reflect the employer's wishes while giving proper scope to the contractor. The contract is unforgiving to the extent that badly assembled Requirements will result in Contractor's Proposals which do not satisfy the employer.
- Because the relationship between employer and contractor's architect is not the close one of client and independent consultant, because the employer will not usually choose the architect and because the architect may be under instructions from the contractor to design down to a price, the quality of design may not be as good as a building produced in the traditional fashion.
- The employer will be involved in additional fees. The design fees which the employer would normally pay to consultants will be included in the contractor's design and build price. The employer will need independent professional advice and, therefore, the employer will have to pay extra for it.

1.3 The architect's role

Architects are said to dislike design and build. There are several reasons advanced for this, including the suggestion that where the client does not appoint the architect, the standard of design will necessarily suffer.

Although the architect will not have the role ascribed under the Standard Building Contract (SBC), the architect cannot be discarded, because someone has to design the building. In addition, the employer will still require independent professional advice in order to use the system to best advantage. So far as the design aspect is concerned, design and build can be extremely flexible.

In order to fully understand the extent of the flexibility, it is useful to consider the extreme situations. At one extreme, the employer may approach a design and build contractor as soon as the intention to build starts to take shape. The contractor, either by means of an in-house architectural department or more commonly by sub-letting the work to an independent architect, takes details of the brief and proceeds through the stages from inception to completion. This is true design and build, where the contractor is responsible for everything from start to finish. It is usual to negotiate the contractor's price, because tendering among a number of contractors is not practicable in this instance.

At the other extreme, the employer may engage a full team of consultants to act in the traditional way in taking the brief and preparing a feasibility report, outline proposals and a detailed design together with a very full specification. Tendering then takes place and the successful tenderer proceeds on the basis that it takes responsibility for completing the detailed design as well as constructing the building. In practice, that will involve the contractor in producing a full set of production information. The employer has little to gain by adopting this system, because once the design team has designed the building, there is every reason for retaining them to deal with inspections and queries during the construction period under a traditional contractual arrangement. In any event, the employer will require some kind of independent advice at this time. One comes across this particular variant quite often and, in some instances, it originates in the employer's intention to proceed down the traditional path and its implementation until rather late in the process. The employer tries to have the best of both worlds – full control over design but with one point responsibility. Such late changes of mind often result in complex design liabilities and many opportunities for claims. In such instances, blame lies at the feet of the employer and the employer's professional advisors.

A common practice somewhere between the two extremes involves the design team in taking the brief, carrying out feasibility studies and preparing outline proposals and a performance specification for tendering purposes. The successful tenderer is responsible for completing the design using its own team, and the employer's team is available to assist the employer with advice throughout the construction period. Some of these variants are shown in diagrammatic form in Fig. 1.1.

Interesting variations are consultant switch and novation. The system requires the employer to appoint a design team in the traditional way and the team takes the employer's brief, prepares feasibility studies and develops proposals to a fairly advanced stage with a performance specification. In consultant switch, tendering takes place on the basis that the successful contractor will enter into a new contract with each of the design consultants. In novation, tendering takes place on the basis that the employer, successful contractor and each consultant will enter into a

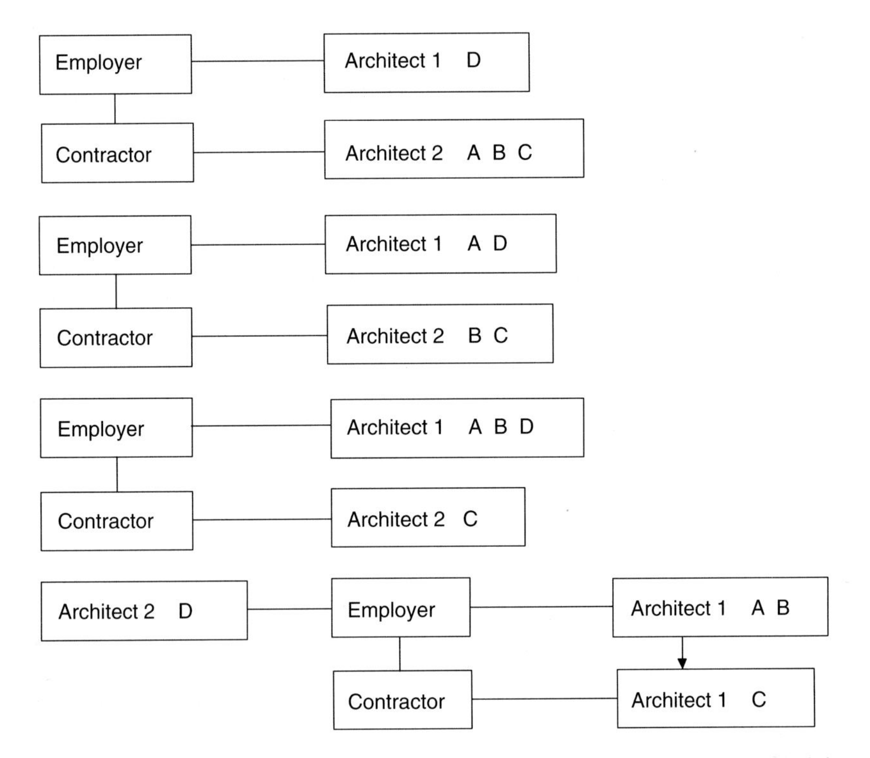

Fig. 1.1 Design and build variants. A = Brief stage; B = outline proposals; C = detailed design stage; D = advice during construction.

novation agreement which effectively removes the employer and substitutes the contractor in the consultant's appointment document. This is supposed to avoid any danger of a design responsibility split between the employer's and the contractor's architects and it is also supposed to ensure a high degree of design continuity. Care must be taken, however, because the design team's duties to the contractor will not be the same as their duties to the employer. Therefore, simply to novate the contracts between design team and employer to the contractor will not work – although it is commonly done. The individual members of the design team and the contractor must have in their contracts with the employer the terms of the contracts between design team and the contractor, together with an undertaking to enter into a future contract on those terms. (For a detailed description of novation see sections 3.4 and 6.1.)

Obviously, once the team have entered into contracts with the contractor, they can no longer give independent advice to the employer. With this system, the employer must either accept that no independent advice will be given after tender stage or, and this is more likely, other consultants must be engaged as necessary to provide the required advice. Problems may arise, for example, if the second architect disagrees with the first architect's design. The architect's design liability in this and other situations is discussed in Chapter 3.

Another important way in which design responsibility can be split is when everything is carried out traditionally, but some element of the building, such as the foundations or a floor, is left for the contractor to design. In general, this is treated as though the element was a miniature design and build contract within the traditional contract framework. If it is thought essential to split off part of the design responsibility in this way, it is crucial that the element is as self-contained as possible otherwise the task of sorting out respective design responsibilities becomes a nightmare.

1.4 *Standard forms available*

A multitude of forms of contract have been used for design and build procurement. It is still all too common to see traditional forms such as the former JCT 98 heavily amended in an attempt, rarely successful, to produce something suitable for design and build. JCT 98 and its successor, SBC, are wholly unsuitable for use as a design and build contract. For many years, much design and build was carried out using contractors' in-house forms, and this is sometimes the case even today, despite the availability of an acceptable standard form of contract.

1.4.1 The Design and Build Contract (DB)

This form was originally published in 1981 (as CD 81) by the Joint Contracts Tribunal (now the Joint Contracts Tribunal Limited) and imposes on the contractor a liability for the design equivalent to that imposed on an architect or other professional designer, i.e. an obligation to use reasonable skill and care in the preparation of the design.

CD 81 gained steadily in popularity and it became the single most commonly used form. It was modelled on the layout and the wording of the Standard Form of Building Contract 1980 (JCT 80, later JCT 98, now SBC), a fact which could be useful for those who were already conversant with JCT 80. There was the problem, however, that the user might not appreciate the degree to which CD 81 differed from JCT 80, often in quite subtle ways but to a significant extent. These differences persist between SBC and DB.

The JCT have produced the *Design and Build Contract Guide* which sets out a brief overview of the contract and brief comments on the various provisions. The principal changes can be seen at a glance. There is a very useful table of destinations at the back of the Guide which helpfully lists the clauses in the previous WCD 98 by clause number and shows the new clause number in DB. It is probably worth getting the Guide for this feature alone.

The criteria for the use of the contract are set out inside the front cover of the contract itself. They indicate that DB is an appropriate contract where:

- Detailed provisions are necessary and Employer's Requirements have been prepared for the contractor.
- The contractor is to complete the design as well as carry out the Works.
- The employer employs an agent to administer the contract.

It is also pointed out that the contract can be used if the Works are to be carried out in sections or whether the employer is private or a local authority.

The DB form of contract is examined in detail in the remainder of this book.

1.4.2 The ACA Form of Building Agreement 1982, Third Edition 1998, (2003) (ACA 3)

This form was first published by the Association of Consultant Architects in 1982. It was subjected to much criticism – a great deal of which was emotional and unjustified – but it was amended substantially in 1984. It was again revised in minor respects in 1990 and in 1998 to comply with the Housing Grants, Construction and Regeneration Act 1996 Part II. The latest revision is dated 2003. A useful *Guide to the ACA Form of Building Agreement* (1982) Third Edition (1998) (2003 revision) is available. There is an edition of ACA 3 which has been specially adapted to the needs of the British Property Federation (BPF) system of building procurement. An important feature of ACA 3 is the provision of standard alternative clauses. Although ACA 3 is basically a traditional form, the proper combination of alternative clauses can produce a design and build variant. It is a relatively simple form with clearly defined divisions of responsibility. It is not a negotiated form like DB; in the case of a dispute, any ambiguity is likely to be construed against the employer who puts it forward. The form may also be classed as the employer's written standard terms of business for the purposes of the Unfair Contract Terms Act 1977. This can affect any clauses which are deemed to be exclusions or restrictions of liability.

The key clause is clause 3.1 in which the contractor warrants that the Works will comply with any performance specification in the contract documents, and that the parts of the Works to be designed by the contractor will be fit for purpose. The design warranty could scarcely be wider and equates the contractor's position with the duty of a seller of goods to supply goods which are reasonably fit for their intended purpose. The contractor is responsible for any failure in the design irrespective of fault and the contractor must maintain design indemnity insurance under clause 6.6. The form can also accommodate small parcels of design by the contractor within a basically architect-designed framework.

Although this form has many virtues, it has not made the impact it deserves, perhaps because some contractors see it as heavily weighted in favour of the employer.

1.4.3 The NEC3 Engineering and Construction Contract (2005) (NEC3)

This form was first published for the Institution of Civil Engineers in 1993. It employs a rather different philosophy to other standard forms and partly as a result it has received some criticism. It was also the subject of fairly unrestrained praise by Sir Michael Latham in his report 'Constructing the Team' (the Latham Report). It is not a specialist design and build form, but it is said to be flexible enough to support design and build as an option, somewhat like the ACA form. The basic principle of this form, which is good, is that there are a number of unchanging core clauses onto which can be grafted any one of six main option clauses (such things as priced contract with activity schedule or cost reimbursable contract, etc.). Other clauses (performance bond, retention, trust fund, etc.) can be added if desired. The contract has a strange numbering system (e.g. 40.6 is a subclause of clause 4) and it is mostly written in the present tense so that it is impossible to be sure which of the actions are intended to be duties and which are optional (powers). An added difficulty is that the authors appear to have eschewed any words which have ever been defined in the courts. Therefore, it is often difficult to be sure what certain words really mean. This is certainly a brave attempt to break the mould, but I cannot recommend it. A very perceptive discussion of this form was written by Donald Valentine and published in *Construction Law Journal*, 1996, vol. 12, p. 305.

An addendum was issued in April 1998 to take into account the Housing Grants, Construction and Regeneration Act 1996. The latest revision is June 2005. The *NEC3 Engineering and Construction Contract Guidance Notes* (June 2005) are available.

1.4.4 Standard Building Contract (SBC), Intermediate Building Contract with contractor's design (ICD), Minor Works Building Contract with contractor's design (MWD)

What used to be published as the Contractor's Designed Portion Supplement for use with JCT 98 is now fully incorporated into the Standard Building Contract (SBC). The Intermediate and Minor Works Building Contracts (IC and MW respectively) are each published in another edition (ICD and MWD), which incorporates provision for a contractor's designed portion (CDP).

The relevant portions of these contracts effectively reproduce the important provisions of DB for a small portion of a contract generally being carried out under traditional contracting procedures. It is intended for the situation when the employer wishes part of the project to be designed by the contractor.

An interesting question arises concerning the architect's obligation to integrate the CDP with the rest of the work (SBC clause 2.2.2, ICD clause 2.1.2, MWD clause 2.1.3). The question is so common that it is worth dealing with here. The position is that the contractor is responsible for the integration of the design (contained in the CDP) with the rest of the Works so far as they can be ascertained by the contractor from the information supplied to it at the date of the contract. If the architect makes no further changes in the Works, the contractor must ensure that its design is properly co-ordinated with the Works as a whole. If, however, the architect issues instructions which change the requirements on which the contractor's design is based, or which change the design of the Works, it is for the architect to give such instructions as may be necessary to achieve a proper integration of either the changed contractor design into the unamended rest of the Works or the contractor design into the changed rest of the Works.

A comparison of DB and ACA 3 clauses is given in Fig. 1.2.

Description	**DB**	**ACA 3**
Definitions and interpretation	1	23
Definitions	1.1	23.2
Reference to clauses etc.	1.2	
Agreement etc. to be read as a whole	1.3	1.3
Headings etc.	1.4	23.2
Reckoning periods of days	1.5	
Contracts (Rights of Third Parties) Act 1999	1.6	
Giving or service of notices etc.	1.7	23.1
Electronic communications	1.8	
Effect of final account and final statement	1.9	
Effect of payments other than payment of final statement	1.10	
Applicable law	1.11	25C
Carrying out the Works	2	1
General obligations	2.1	1.1, 1.2
Materials, goods and workmanship	2.2	3.5

Fig. 1.2 Comparison of clauses in standard form of contract DB with ACA 3.

Date of possession – progress	2.3	11.1
Deferment of possession	2.4	
Early use by employer	2.5	
Work not forming part of the contract	2.6	10
Contract documents	2.7	2.1
Construction information	2.8	2.2
Site boundaries	2.9	
Divergence in Employer's Requirements and definition of site boundary	2.10	
Preparation of Employer's Requirements	2.11	
Employer's Requirements – inadequacy	2.12	
Notification of discrepancies etc.	2.13	1.5
Discrepancies in documents	2.14	1.5
Divergences from statutory requirements	2.15	1.6
Emergency compliance with statutory requirements	2.16	
Design work – liabilities and limitations	2.17	3.1
Fees or charges legally demandable	2.18	
Royalties and patent rights – contractor's indemnity	2.19	
Patent rights – instructions	2.20	
Materials and goods – on site	2.21	6.1
Materials and goods – off site	2.22	6.1
Related definitions and interpretation	2.23	
Notice by contractor of delay to progress	2.24	
Fixing completion date	2.25	11.6, 11.7
Relevant events	2.26	11.5
Practical completion	2.27	12.1

Fig. 1.2 *Continued*

Non-completion notice	2.28	11.2
Payment or allowance of liquidated damages	2.29	11.3
Contractor's consent	2.30	13.1
Practical completion date	2.31	13.2
Defects etc. – relevant part	2.32	13.2
Insurance – relevant part	2.33	13.2
Liquidated damages – relevant part	2.34	13.3
Schedules of defects and instructions	2.35	12.2
Notice of completion of making good	2.36	
As-built drawings	2.37	
Copyright and use	2.38	
Control of the Works	3	
Access for employer's agent	3.1	4
Person-in-charge	3.2	5
Consent to sub-letting	3.3	9.2
Conditions of sub-letting	3.4	
Compliance with instructions	3.5	8.1
Non-compliance with instructions	3.6	
Instructions to be in writing	3.7	8.3
Provisions empowering instructions	3.8	8.4
Instructions requiring changes	3.9	8.1(e)
Postponement of work	3.10	11.8
Instructions on provisional sums	3.11	16.7
Inspection – tests	3.12	8.1(c)
Work not in accordance with the contract	3.13	8.1(a)
Workmanship not in accordance with the contract	3.14	

Fig. 1.2 *Continued*

Effect of find of antiquities	3.15	14.1
Instructions on antiquities	3.16	14.2
Loss and expense arising	3.17	7.1
Undertakings to comply with the Construction (Design and Management) Regulations 2007 (CDM)	3.18	26.1
Appointment of successors	3.19	26.4
Payment	4	16
Adjustment only under the conditions	4.1	15.1
Items included in adjustments	4.2	16.2
Taking adjustments into account	4.3	
VAT	4.4	16.8
Construction Industry Scheme (CIS)	4.5	24
Advance payment	4.6	
Issue of interim payments	4.7	16.1
Amounts due in interim payments	4.8	16.3
Application by contractor	4.9	16.1
Interim payments	4.10	16.3
Contractor's right of suspension	4.11	
Final account and final statement – submission and payment	4.12	19.1
Ascertainment – alternative A	4.13	16.2A
Ascertainment – alternative B	4.14	16.2B
Off-site materials and goods	4.15	6.1
Rules on treatment of retention	4.16	16.4
Retention – rules for ascertainment	4.17	16.5
Fluctuations – choice of provisions	4.18	
Matters materially affecting regular progress	4.19	7

Fig. 1.2 *Continued*

Relevant matters	4.20	7.1
Amounts ascertained – addition to contract sum	4.21	7.4
Reservation of contractor's rights and remedies	4.22	
Changes	5	17
Definition of changes	5.1	8.2
Valuation of changes and provisional sum work	5.2	17.5
Giving effect to valuations, agreements, etc.	5.3	
Measurable work	5.4	
Daywork	5.5	
Change of conditions for other work	5.6	
Additional provisions	5.7	
Injury, damage and insurance	6	6
Liability of contractor – personal injury or death	6.1	6.3(a)
Liability of contractor – injury or damage to property	6.2	6.3(b)
Injury or damage to property – Works and site materials excluded	6.3	
Contractor's insurance of its liability	6.4	6.3
Contractor's insurance of liability of employer	6.5	6.5
Excepted risks	6.6	
Insurance options	6.7	6.4
Related definitions	6.8	
Sub-contractors – specified perils cover under joint names all risks policies	6.9	
Terrorism cover – non-availability – employer's options	6.10	
Obligation to insure	6.11	6.6
Increased cost and non-availability	6.12	

Fig. 1.2 *Continued*

Application of clauses	6.13	
Compliance with Joint Fire Code	6.14	
Breach of Joint Fire Code – remedial measures	6.15	
Joint Fire Code – amendments/revisions	6.16	
Assignment, third party rights and collateral warranties	7	
General	7.1	9.1
Rights of enforcement	7.2	
References	7.3	
Notices	7.4	
Execution of collateral warranties	7.5	
Rights for purchasers and tenants	7A	
Rights for a funder	7B	
Contractor's warranties – purchasers and tenants	7C	
Contractor's warranty – funder	7D	
Sub-contractor's warranties	7E	
Termination	8	20
Meaning of insolvency	8.1	20.3
Notices under section 8	8.2	23.1
Other rights, reinstatement	8.3	22.5
Default by contractor	8.4	20.1
Insolvency of contractor	8.5	20.3
Corruption	8.6	
Consequences of termination under clauses 8.4–8.6	8.7	22.1

Fig. 1.2 *Continued*

Employer's decision not to complete the Works	8.8	
Default by employer	8.9	20.2
Insolvency of employer	8.10	20.3
Termination by either party	8.11	21
Consequences of termination under clauses 8.9–8.11 etc.	8.12	22.2
Settlement of disputes	9	25
Mediation	9.1	
Adjudication	9.2	25B
Conduct of arbitration	9.3	
Notice of reference to arbitration	9.4	25.8, 25.9
Powers of arbitrator	9.5	25.10
Effect of award	9.6	
Appeal – questions of law	9.7	
Arbitration Act 1996	9.8	
Contractor's design submission procedure	Schedule 1	2.2, 2.3, 2.4, 2.5
Supplemental provisions	Schedule 2	
Insurance options	Schedule 3	6.4
Code of practice	Schedule 4	
Third party rights	Schedule 5	
Forms of bonds	Schedule 6	
Fluctuations options	Schedule 7	18

Fig. 1.2 *Continued*

1.5 Tendering procedures

Although the employer may use any method of choice to invite and accept tenders, it is wise to follow an established procedure if the employer is to obtain the right contractor providing the right building at the right price. This is true of tendering for any kind of building procurement system, but it is especially true where design and build is concerned. The employer can be vulnerable if the documentation and procedures are not properly completed, and the tenderers can be put to much abortive work. It is unfortunately frequently found that employers who put little effort into the preliminary stages are faced with additional costs during the construction period in order to get what they actually want. Design and build is not a way to avoid making important decisions about building. The success of the finished project will reflect the amount of time the employer is willing to devote to it before tender stage.

The Construction Industry Board (CIB) has produced a *Code of Practice for the Selection of Main Contractors* (1997), which replaces the Code produced by the National Joint Consultative Committee (NJCC). It will repay careful study. Indeed, it is required reading for anyone about to embark on design and build, whether employer, contractor or employer's agent. The Code recognises that selection may be by competitive tendering, negotiation, partnering or a joint venture arrangement, but focuses on competitive tendering on the basis of single stage or two stage procedures. Tenders over the currently specified value in the public sector must be invited in accordance with the appropriate EEC directives. The number of firms invited to tender should be severely restricted, probably to three or four, depending on the type and size of building. It should be remembered that tendering for design and build work involves all tenderers in high cost. For this reason, the list of possible tenderers should be prepared with care. The following must be borne in mind:

- The firm's financial standing
- Recent experience of designing and building the same kind of building
- Whether design will be in-house, and if not, by whom
- Quality of work produced
- General experience and reputation
- Adequacy of management
- Health and safety record
- Adequacy of capacity.

Each firm on the short list should be sent a preliminary enquiry to discover if it is willing to tender. The enquiry should contain the following information if known:

- Job title
- Location of site including plan
- Employer
- Employer's agent
- Availability of and restrictions on services
- General description of work
- Approximate cost range

- Number of tenderers proposed
- Tendering method: single stage or two stage
- Contractor's involvement in planning procedures
- Whether a conservation area
- Form of contract and amendments
- Variable contract details
- Whether Defective Premises Act 1972 applies
- Any limit on the contractor's liability
- Nature and extent of contractor's design input
- Simple contract or deed
- Anticipated date for possession
- Contract period
- Anticipated date for dispatch of tender documents
- Length of tender period
- Length of time for which tender must remain open for acceptance
- Liquidated damages
- Bond
- Special conditions
- Consideration of alternative tenders
- Basis of assessment of tenders
- That tenders will be considered in accordance with the Code of Practice for the Selection of Main Contractors.

It is of great importance that the preliminary enquiry states to what extent the acceptance will depend on factors other than price. To aid in assessment, the employer must state exactly what is required to be submitted with the contractor's tender. Contractors who respond positively should be interviewed at briefing sessions to reduce the choice to the predetermined number of tenderers. If any prospective tenderer has to withdraw, it should give notice before the issue of tender documents.

Note that:

- The tender documents should be despatched on the stated date.
- All tenders should be submitted on the same basis.
- The tender period will be not less than 12 weeks.
- The employer should consider the scope for alternative offers.

Tenderers wanting clarification should notify the employer, who must inform all tenderers of any decisions. Under English law, a tender may be withdrawn at any time before it is accepted even if stated to be open for a number of weeks. The way to ensure that a tender is not withdrawn before the period expires is for the employer to pay a sum of money to the contractor in consideration of the tender being kept open for a specified period. This creates a binding contract. The sum of money is usually nominal and the matter is usually covered in a paragraph in the tender where the contractor confirms 'that, in consideration of a payment of £1 by the employer (receipt of which is hereby acknowledged)' the contractor 'agrees to keep the tender open for a period of . . . weeks from the date hereof'.

Tenders which do not comply with the conditions should be rejected. Unsuccessful tenderers should be informed as soon as a tender has been accepted or a tenderer has been selected to proceed to the second stage, as appropriate. If errors are found in the priced document, the employer must take the appropriate steps as set out in the invitation to tender. That is either:

- If the overall price is stated in the tender enquiry to be dominant, give the contractor the opportunity to confirm or withdraw its tender; or
- If the pricing document is stated in the tender enquiry to be dominant, ask the contractor for an amended tender price to accord with the rates given by the contractor.

Where single stage tendering is involved, there is still scope for negotiation if the preferred tender is too high. Where two stage tendering is adopted, there must be a clear definition of the following matters:

- Grounds for withdrawal from the second stage.
- Entitlement to costs and method of ascertainment if second stage negotiations are not concluded to mutual satisfaction.
- Reimbursement for any work done on site if second stage procedures are abortive.

It is not suggested that a first stage tender should be accepted. There can be no true acceptance at that stage because of the need to leave open the second stage negotiation procedure, and neither party wishes to have a concluded contract at that stage. It is probably better to avoid anything that appears to be an acceptance and the employer should simply notify the successful tenderer of the intention to continue negotiations in the hope of achieving a mutually satisfactory outcome. Sometimes, employer and contractor enter into a pre-construction services agreement to cover the period between receipt of preliminary proposals and finalising the Contractor's Proposals and contract sum. If such an agreement is contemplated, it should be drawn up with care. It is important that the agreement makes clear the basis of the negotiations and that, if the contractor does not secure the project, it will have no redress against the employer in terms of loss of profit and other losses. Whether the employer is prepared to pay the contractor during this period or whether securing the contract is considered to be reward enough is a matter for individual circumstances.

The second stage is really the finalisation of the Contractor's Proposals. Where contractors have been notified that specific conditions will apply to the tendering process, the employer must strictly adhere to such conditions. Any failure in this respect might entitle the contractor to recover damages, unless the employer is protected against liability by means of a suitably-worded clause in the tender documentation. By setting out terms for tendering, the employer is making an offer, in a limited way, that if a tenderer submits a tender in response, the employer will proceed according to such terms. A contract is formed and breach of its terms will enable the other party to recover damages for proven loss. Such damages would generally amount to the cost of preparing the tender, which could be a substantial sum where design and build is concerned. In appropriate circumstances, it is conceivable (but debatable) that a tenderer whose tender was not properly considered could claim the loss of the profit it would have made had it been properly awarded the contract.

These propositions derive from the Court of Appeal decision in *Blackpool & Fylde Aero Club* v. *Blackpool Borough Council* (1990), where the defendants invited tenders for a concession. The tender document stated that the defendants did not bind themselves to accept 'all or any part of any tender' and also that 'no tender which is received after the last date and time specified will be admitted for consideration'. Tenders had to be received by the Council 'not later than 12 o'clock noon on Thursday 17 March 1983'. At 11 am on 17 March the plaintiffs' representative put their tender into a letterbox at the Town Hall. A notice on the box stated that the box was emptied daily at 12 o'clock noon. In the event, the plaintiffs' tender was not taken from the box until 18 March, and was excluded on the grounds that it was too late. The concession was awarded to another tender and the plaintiffs sued alleging breach of contract. The Court of Appeal upheld the claim, holding that there was a contractual obligation to consider any tender properly submitted in accordance with the stipulated and detailed conditions of tendering. In effect, a contract was implied.

However, the *Blackpool* case was distinguished by a differently constituted Court of Appeal in *Fairclough Building Ltd* v. *Port Talbot Borough Council* (1992) where it was held that, under the normal tendering process, a tenderer has no cause of action where its tender is rejected but has been given *some* consideration and the recipient of the tender has acted reasonably.

In that case, the Council decided to have a new Civic Centre constructed, and Fairclough applied to be included on the selective tendering list. Their application was successful and subsequently they were invited to tender. The wife of one of Fairclough's directors (whose name was on the company's letterheads) was employed as an architect by the Council and very properly disclosed her 'interest' under the Local Government Act 1972. In fact the relationship was already known to the Borough Engineer, but as a result Fairclough were removed from the tender list, although it was said that the 'decision is not intended to reflect any doubts whatsoever upon the integrity of your company or the individual member of staff'.

Fairclough considered that there was a breach of contract and, on appeal, relied on the *Blackpool* case. The Court of Appeal ruled against Fairclough and held that the Council had fulfilled its obligation by giving some consideration to the tender and had acted reasonably. *Blackpool and Fylde Aero Club* v. *Blackpool Borough Council* (1990) was distinguished on somewhat slender grounds.

The case of *Pratt Contractors Ltd* v. *Transit New Zealand* (2003) was an appeal from the Court of Appeal in New Zealand to the Judicial Committee of the Privy Council of the House of Lords in Westminster. The facts are that a state highway contract was put out to competitive tender. Pratt submitted tenders, but it was unsuccessful despite being the lowest. Pratt said, and Transit agreed, that an invitation to tender together with the submission of the tender constituted a contract under which Transit was obliged to carry out the procedure for choosing a tenderer which it had set out in the invitation. In addition, it should act fairly and in good faith. Unfortunately, the parties could not agree what that amounted to in practice.

Pratt alleged that Transit had been in breach of its obligations and sought damages. There were two ways in which Transit could have assessed the tenders in accordance with the obligatory procedure. One was the straightforward lowest price tender. The other, which Transit employed, was a formula which allotted different marks for various criteria.

The tender panel excluded Pratt from consideration because it failed some of the criteria. The panel gave detailed reasons in their subsequent report. Transit was reluctant to take this advice and eventually it re-advertised. This time Pratt passed all the criteria, but another contractor had a higher score and it was awarded the contract.

The Privy Council decided that, although Transit had a duty to carry out the procedures properly, the duty to act fairly and in good faith amounted to no more than that the panel should express their views honestly. Transit had no obligation to appoint, as panel members, people without any existing views about the tenderers. Indeed, people with the requisite experience to serve on the panel were just the sort of people to have already formed views about the contractors. Consequently, Pratt's claim was dismissed.

These cases, when read together, indicate the employer's duty to do what it undertook to do in inviting tenders. The cases do not appear to impose any heavier duty than that.

Chapter 2
Contract Documents

2.1 *The documents*

The contract documents can be any documents which are evidence of the contract. They are agreed by the parties to the contract and signed. It is important that each document is signed by both parties and dated. To avoid any doubt, it is customary for each document to be endorsed: 'This is one of the contract documents referred to in the Agreement dated . . .'. DB, unlike its predecessor WCD 98, now expressly defines the contract documents. These are:

- Agreement
- Conditions
- Employer's Requirements
- Contractor's Proposals
- Contract Sum Analysis.

The contents of DB are arranged as follows:

Articles of Agreement
Recitals
Articles
Contract particulars
Attestation
Conditions
1. Definitions and interpretation
2. Carrying out the Works
3. Control of the Works
4. Payment
5. Changes
6. Injury, damage and insurance
7. Assignment, third party rights and collateral warranties
8. Termination
9. Settlement of disputes

Schedules
1. Contractor's design submission procedure
2. Supplemental Provisions
3. Insurance options
4. Code of practice
5. Third party rights

6. Forms of bonds
7. Fluctuations options

For the purposes of this book the layout of the printed form has not been followed; rather, what seems to be a more logical arrangement has been adopted, dealing with the form on a topic basis and making reference to appropriate clauses, wherever they may be located.

2.2 *Completing the form*

Care must be taken in completing the form. This task is normally undertaken by the employer's professional advisor, i.e. whoever is the employer's agent under the provisions of article 3. Sometimes it is considered necessary to make amendments to the clauses in the printed form. If possible, such amendments should be avoided, but if it is not possible, each amendment or deletion should be clearly made in the appropriate place on the form and each party should initial, preferably at the beginning and end of the amendment especially where a deletion has been carried out.

Articles of Agreement

The date is always left blank until the form is executed by the parties. The names and addresses of the employer and the contractor must be inserted in the space provided. Where limited companies are involved, it is sensible to insert the company registration number in brackets after the company name so that there is no possible chance of confusion in cases where companies change or even exchange names. It is good to see that the contract now provides a space for that purpose.

The first recital is important. The description of the work must be entered with care, because among other things it can affect the operation of the variation clause (clause 5). The contract sum is to be inserted in article 2. It should be noted that the amount of the contract sum will never change. Where the contract permits or requires amounts to be added to or deducted from the contract sum, the result is referred to as the 'adjusted contract sum'. The name of the employer's agent must be inserted in article 3 unless the employer has unwisely decided against the employment of an agent. Normally, the agent's name will be the name of a firm.

The fifth recital is to be deleted if the Works are not divided into sections.

Article 5 must be completed to indicate the identity of the CDM co-ordinator under the CDM Regulations. The default position is that the contractor takes on that role. If another person is to take the role, the name and address must be inserted. Where the contractor takes the role, it is highly likely that another CDM co-ordinator will be appointed for the early stages of the project, only changing to the contractor when tenders are accepted.

Article 6 must be completed with the identity of the principal contractor. Again, the default position is that the contractor will take that role. In the unlikely event that another person is to be appointed, the details must be inserted. If the project is not notifiable, both articles 5 and 6 should be deleted.

Contract particulars

This is an important variable part of the contract. It is necessary to consider it with great care. Errors in this part tend to have severe financial repercussions. It should be completed as follows:

Part 1: General

Fourth recital and clause 4.5: For the purposes of the Construction Industry Scheme (CIS), it is necessary to state whether the employer at the base date is or is not a 'contractor'.

Fifth recital: If the Works are to be divided into sections, they are to be described here. This is usually done by reference to an attached drawing. A note to this entry assumes that any sections will be described in the Employer's Requirements. That would be good practice. Otherwise, the document concerned should be stated here, attached firmly to the contract and signed and dated by the parties.

Article 4: There are three entries against this side heading. The Employer's Requirements or the Contractor's Proposals or the Contract Sum Analysis as appropriate should be identified here. Each document should be signed and dated by the parties. Depending on the size it should be either firmly attached to the contract or identified as one of the contract documents. See section 2.1 for a suitable endorsement.

Article 8: This must be completed to indicate whether arbitration or legal proceedings will be the final dispute resolution process. If arbitration is the preferred choice, the words 'do not apply' must be deleted so that article 8 and clauses 9.3 to 9.8 do apply. If nothing is deleted, the default position is legal proceedings. It is a great pity that previous practice on this point has been reversed, possibly causing many people to inadvertently choose legal proceedings when believing that arbitration was still the default process. A footnote ([10]) refers the parties to the *Guide* for the factors to be taken into account when choosing between arbitration or legal proceedings. The advice in the *Guide* appears to be substantially in favour of legal proceedings, but not all factors are considered. A full consideration of all the factors can be found in section 13.1.

1.1 Base date: This is an important date and referred to in several of the contract clauses. It used to be the date of tender, but that was often uncertain, because the period for tendering was often extended or even postponed for long periods. Any date can be chosen as the base date, but it is usually a date corresponding to the date tenders are received. Clauses 2.26.11 (exercise by the UK government of statutory power after the base date), 4.4 (position on supply of goods and services which become exempt from VAT after base date), 5.5 (daywork definitions current at base date) and 6.16 (amendment of Joint Fire Code after base date) are examples in the contract of references to the base date.

1.1 CDM planning period: This is the time which must be allocated prior to commencement of construction so that various parties can do the CDM planning and preparation. It should be noted that the period may pre-date that of possession.

1.1 Date for completion: This must be inserted if there are no sections. If the Works are divided into sections, the first entry must be ignored or struck through and the relevant completion date must be inserted against each section. There is space for only three sections. Therefore, if there are more than three sections or if the sections need to be described in more detail than by numbers, reference should be made here to a separate sheet containing the information which should be signed and dated by each party and firmly attached to the contract.

1.7 Addresses for service of notices: The relevant addresses and fax numbers for employer and contractor must be inserted. There is a note to the effect that if nothing is inserted, the addresses shown at the beginning of the contract will be used until the parties agree otherwise. This provision is made subject to clause 1.7.2, that if there is no current agreed address, the notice will be effectively served by sending it by post to the last known principal business address, registered or principal office. Since the object of inserting the note is presumably to ensure that there is always by default a current agreed address, it is difficult to see when clause 1.7.2 will apply.

1.8 Electronic communications: If this entry is not completed, it states that all communications must be in writing. Once again, if there is insufficient space, reference should be made here to a separate sheet containing the information, which should be signed and dated by each party and firmly attached to the contract. It is not only the type of correspondence which must be stipulated, for example, issue of drawings or general correspondence, queries and so on, but also the format in which such communications are to be made. For example, all text attachments might be required in WordPerfect or Word. Electronic communications may not be used where the contract expressly states that communication must be by a particular method (clause 1.7); this is by no means as clear as it could be (see section 2.9).

2.3 Date of possession: This must be inserted if there are no sections. If the Works are divided into sections, the first entry must be ignored or struck through and the relevant date of possession must be inserted against each section. There is space for only three sections. Therefore, if there are more than three sections or if the sections need to be described in more detail than by numbers, reference should be made here to a separate sheet containing the information, which should be signed and dated by each party and firmly attached to the contract.

2.4 and 2.26.3 Deferment of possession: If the employer wishes to have power to defer possession of the site, the entry must be completed to show that clause 2.4 applies. It is wise to make clause 2.4 apply even if there appears to be nothing to prevent the contractor taking possession on the due date. The unexpected and the unlikely usually occur if no preparations are made. The maximum period of deferment is stated as 6 weeks and if a lesser period is needed it should be inserted. It is always good practice to insert a period, even if it is the full 6 weeks. It is not immediately obvious why an employer would wish to limit the period to less than 6 weeks in any event. If the Works are divided into sections, the first entry must be ignored or struck through and the entry must be completed to show that clause 2.4 applies. The relevant period of deferment must be inserted against each section. There is space for only three sections. Therefore, if there are more than three sections or if the sections need to be described in more detail than by numbers, reference should be made here to a separate sheet containing the information, which should be signed and dated by each party and firmly attached to the contract.

2.17.3 Limit of contractor's liability for loss of use: Although contractors will understandably prefer this entry to be completed with a 'nil' figure, employers will usually insert the words 'no limit'. It is possible that the insertion of a limit on the contractor's liability may result in a lower tender, but it may be sensible to insert 'no limit' at tender stage and see whether the inclusion of a limit will result in a reduction of the tender sum. It seems unlikely (see section 3.2).

2.29.2 Liquidated damages: The rate of liquidated damages is to be inserted, and the period. For example: '£100.00 per week'. It is important that the figures are very clearly written so that there is no possibility of mistake. It is surprising how many contracts one sees where the amount of liquidated damages is scribbled in almost illegibly. If the amount is not clear, liquidated damages cannot be imposed. In view of the huge sums which can accrue in such damages, it is worth taking great care.

It used to be very common to see the words ' per week or part thereof' inserted. It is less common now, but it still occurs. The words mean that the amount of liquidated damages is recoverable for every week and for every part of a week, even if that part is just a few minutes. At first sight, such a provision is sufficient to turn the liquidated damages into a penalty on the basis that a sum which may be a genuine pre-estimate of expected loss for 1 week must be excessive for a part of that week. Of course, penalties are not enforceable. It may be that the sum can be shown to be correct, for example, it may represent rental which becomes due at the beginning of each week. What most people mean when they use the phrase is 'per week or pro rata for a part of a week', indicating that the damages are to be proportioned for part of a week. Even these words must be used with care and they should not be used at all if it can be demonstrated that a differing loss would be incurred during certain parts of a week, for example, during the weekend. It is possible to insert a more sophisticated arrangement for damages, but this should be done with proper advice. Section 8.4 deals with liquidated damages in more detail.

If the Works are divided into sections, the first entry must be ignored or struck through. The relevant amount of liquidated damages must be inserted against each section. Care must be taken to ensure that the sum for liquidated damages for each section properly reflects the genuine pre-estimate of loss for that particular section. Although, where each section is identical, such as a number of houses, the damages may be identical, where the sections vary in size, value or type it is likely that the amount of damages will vary also. There is space for only three sections. Therefore, if there are more than three sections or if the sections need to be described in more detail than by numbers, reference should be made here to a separate sheet containing the information, which should be signed and dated by each party and firmly attached to the contract.

2.34 Section sums: Although included in the definitions (clause 1.1), the definition for 'Section Sum' leaves much to be desired. Indeed, it could hardly be less helpful when, with very little effort, it could have been clearly defined so as to be understandable. The current definition simply refers the reader to clause 2.34 and the contract particulars. Looking at clause 2.34, it appears that the section sum is a part of the contract sum allocated to a particular section. Therefore, if the contract sum is £500,000 and there are three sections, the relevant section sums might be; 1: £150,000, 2: £300,000, 3: £50,000, representing the value of the work in each section.

In order to assist this process, it is sensible for the Employer's Requirements to ask for the Contract Sum Analysis to be split in this way.

2.35 Rectification period: This used to be called the 'Defects Liability Period'. It is suspected that the title has been changed to avoid the confusion which often arose because the contractor (and sometimes the architect) used to mistakenly think that the end of the period signalled the end of the contractor's liability. The contractor has the right to re-enter the site to make good any defects which become visible during this period. If no period is inserted, the default period is 6 months. It is common, although quite wrong, to see separate periods inserted for building work and for services. There is no provision in the contract for separate notices of making good and if a longer period is required for services, that should be the period inserted for all the work. If the Works are divided into sections, the first entry must be ignored or struck through and the relevant period must be inserted against each section. It is likely that the periods will be the same for each section. There is space for only three sections. Therefore, if there are more than three sections or if the sections need to be described in more detail than by numbers, reference should be made here to a separate sheet containing the information, which should be signed and dated by each party and firmly attached to the contract.

4.6 Advance payment: This provision is stated not to apply to local authorities. If the employer intends to make an advance payment under this clause, it should be stated to apply by deleting the alternative. The figure is to be inserted as a sum of money or as a percentage. Whichever is chosen, delete the alternative. The date on which the sum is to be paid should be inserted, together with the way in which the amount is to be reimbursed, by inserting a series of amounts and times. For example, the first repayment may be at the third application for payment followed by a similar amount at the time of each successive application until the full amount is repaid. If more space is needed, reference should be made here to a separate sheet containing the information, which should be signed and dated by each party and firmly attached to the contract. If a bond is required for the advance payment, and it would seem foolish not to require one, it should be so stated by deleting the alternative. If neither option is deleted, a bond will be required.

4.7 Method of payment: The options are stage payments or periodic payments. The option not required must be deleted. If stage payments are required, the table should be completed to show the stages and the cumulative values. Note that the stages refer to stages of the building (ground floor slab, first floor joists, etc.) not to periods of time. If there is insufficient space, a separate sheet containing the information should be signed and dated by each party and firmly attached to the contract. The appropriate part of the heading should be deleted. If periodic payments are required, the date of the first application must be inserted. A footnote ([16]) gives further advice if applications are to be on the last day of each month. If nothing is inserted, the default position is that the first application must be made within a month of the date of possession.

4.15.4 Uniquely identified listed items: If a bond is not required, this entry should be deleted. If it is required, insert the amount.

4.15.5 Not uniquely identified listed items: If the clause does not apply, delete the entry. If it does apply, insert the amount.

4.17.1 Retention percentage: The rate should be inserted. If no rate is inserted, it will be 3%.

4.18 and schedule 7 Fluctuations options: The fluctuations clauses are in schedule 7. Two of the three options should be deleted. If no deletions are made, option A applies. Option A is contribution, levy and tax fluctuations, option B is labour and materials cost and tax fluctuations and option C is formula adjustment. Option A normally leads to the least adjustment. If a percentage addition is to be made to the amounts, it should be inserted. Where formula rules apply, the base month must be inserted for rule 3. Where the employer is a local authority the non-adjustable element as a percentage is to be inserted for rule 3. Delete the options to indicate whether, under rules 10 and 30(i), part I or part II of the formula rules apply depending on whether work category method or work group method applies. If neither is deleted, part I will apply.

6.4.1.2 Contractor's insurance: Insert the amount of cover required.

6.5.1 Insurance – liability of the employer: If the Employer's Requirements state that this insurance is required, the minimum amount of indemnity must be inserted. It should be noted that it is for any one occurrence or series of occurrences arising out of one event. If the indemnity is to be for an aggregate amount, an amendment should be made. As in the case of all insurance matters, the employer should obtain special advice.

6.7 and schedule 3 Insurance of the Works: Two of the three options should be deleted. Option A is for new works if the contractor insures, option B is for new works if the employer insures and option C is for work or extensions to existing structures where the employer insures.

6.7 and schedule 3 Insurance option A (paragraphs A.1 and A.3), B (paragraph B.1) or C (paragraph C.2) Percentage for professional fees: The appropriate percentage should be inserted here. If nothing is inserted, the default position is 15%.

6.7 and schedule 3 Insurance option A (paragraph A.3) Annual renewal date for insurance: This is the date that the contractor must provide if it is going to rely on its annual insurance to satisfy this clause.

6.11 Professional indemnity insurance: This is a new provision. The amount of insurance required must be inserted. If nothing is inserted, the default position is that no insurance is required. This will be an unusual situation. In virtually every case an amount will be inserted and one of the two starred items should be deleted to indicate whether it relates to a series of claims arising from one event or an aggregate amount for a period. The employer must obtain specialist advice on the most suitable insurance for the project. If neither option is deleted, the amount will be the aggregate. An amount must be inserted for the level of cover for pollution or contamination. If nothing is inserted, the level will be the full amount of indemnity cover. The period for expiry of the cover should be inserted. Usually, if the contract is executed under hand, the period will be inserted as 6 years. If the contract is a deed, the period is usually inserted as 12 years, to accord with the limitation period in each instance. It may be sensible to make the periods either 7 or 13 years, respectively, to avoid any shortfall in time.

6.13 Joint Fire Code: Delete to indicate whether the code applies. Either 'Yes' or 'No' must be deleted to indicate whether the Works are specified as a 'Large Project' by the insurer. If the insurance is under option A, the information for these entries should be obtained from the contractor.

6.16 Joint Fire Code – amendments: Delete as appropriate to show whether the employer or the contractor is to bear the cost. If no deletions are made, the cost is to be borne by the contractor.

7.2 Assignment of rights: Delete in the first entry to show whether clause 7.2, giving the right, is to apply. If no deletion, the right will apply. Where the Works are divided in sections, rights apply to each section. If the rights are not to apply to any section, delete the second entry. If rights are to apply merely to some of the sections, state the sections involved. This may be better done by reference to a separate sheet containing the information, which should be signed and dated by each party and firmly attached to the contract.

8.9.2 Period of suspension: Insert the period for which the carrying out of substantially the whole of the Works must be suspended under this clause before the contractor is entitled to issue a 14-day notice prior to termination. If nothing is entered, the period is 2 months.

8.11.1.1– 8.11.1.6 Period of suspension: Insert the period for which the carrying out of substantially the whole of the Works must be suspended under one of these clauses before either party is entitled to issue a 7-day notice prior to termination.

9.2.1 Adjudication: If the parties are agreed on the name of the adjudicator and he or she has consented, it should be inserted. Whether or not an adjudicator has been named, the nominator should be chosen by deleting all except one of the listed bodies. If no nominator is chosen, the party requiring adjudication is free to choose any of the listed bodies.

9.4.1 Arbitration: An appointor should be chosen by deleting all except one of the list of appointors. If no appointor is chosen, the appointor will be the President or Vice-President of the Royal Institute of British Architects.

Schedule 2 Supplemental conditions: The first thing to note is that the correct title on schedule 2 is 'Supplemental Provisions'. The use of 'Conditions' is probably a mistake. In completing the particulars it is suggested that the word 'Conditions' is deleted and 'Provisions' inserted in lieu. Although there is probably very little doubt about the text being referred to in any event, it is wise to be consistent and it avoids disputes. A deletion should be made to show whether these provisions apply. Care should be taken in deciding, because some of these provisions effectively override the usual contract clauses. It is particularly important to complete this entry, because there is no default position.

Schedule 2 (paragraph 1.1) Site manager: A deletion should be made to show whether paragraph 1 is to apply. It is clear from paragraph 1 of the schedule that it does not apply unless so stated here. Therefore, if no deletion is made, the default position is that paragraph 1 does not apply.

Part 2: Third party rights and collateral warranties

This is a particularly difficult part of the contract particulars to complete correctly. There are many notes and footnotes, but they must all be read with great care. The first thing to note is that this part need only be completed if third party rights or warranties are required from the contractor or sub-contractors. If so, there is a choice. The particulars (A) to (D) in this part may be completed or the information may be entered on a separate sheet or sheets and identified at the beginning of this part. The sheets should be signed and dated by the parties and firmly attached to the contract. If it is decided to complete part 2 instead, it must be done as follows (obviously, the same information must be given on the attached sheet):

(A) Identity of purchasers or tenants to receive third party rights or warranties: The name, class or description of the purchasers and/or tenants who are to receive the rights is to be listed in the left-hand column. The correct names should be inserted if known. If not known at this stage, some description which will sufficiently identify the beneficiary should be given, for example: 'All first tenants' or 'purchasers of blocks A, B, C, etc.'. The middle column must contain a description of the part of the Works concerned and which is to be purchased or leased. The right-hand column must be completed to state whether clause 7A (third party rights) or 7C (collateral warranties) is to apply.

(B) Purchaser and tenant rights from the contractor: This refers to the third party rights set out in schedule 5, part 1, or the JCT collateral warranty CWa/P&T. The paragraph numbers of the first are identical to the clause numbers of the second. The entries must be completed so as to show whether paragraph/clause 1.1.2 is to apply. If it does apply, the contractor will be liable for any purchaser's or tenant's losses, other than set out in paragraph/clause 1.1.1, up to the figure which must be inserted in this part (B) as the maximum liability. The type of liability, whether in respect of each breach or aggregate, must also be shown. The default position is that if no figure is inserted or no type of liability is shown, 1.1.2 will not apply.

(C) Identity of funder: The funder who is to receive the rights must be identified by name, class or description. If the name is known, that should be inserted. Otherwise, the description must be quite unambiguous, for example, 'The bank/finance house/organisation providing funding to the employer for the execution of the Works'. If this entry is not completed, the contractor will not be required to give any rights to the funder.

(D) Funder rights from the contractor: The type of right must be shown by deleting one of the options to leave either third party rights or collateral warranty. A period must be inserted within which the contractor may not issue a notice prior to termination nor informing the employer that the contract is being treated as repudiated. If no period is inserted, the default period of 7 days will apply.

(E) Collateral warranties from sub-contractors: If warranties are required from sub-contractors, the particulars in this part may be completed or the information may be entered on a separate sheet or sheets and identified at the beginning of this part. The sheets should be signed and dated by the parties and firmly attached to the contract. Consultants employed by the contractor may also be included in this part. If it is decided to complete part (E) instead, it must be done as follows (obviously, the same information must be given on the attached sheet):

The sub-contractors or consultants should be listed in the left-hand column. If it is not possible to name any of them, an unambiguous description must be given, for example, 'mechanical services sub-contractor' or 'structural engineering consultant'. In the middle column state the type of warranty required: SCWa/F, SCWa/PET or SCWa/E. If professional indemnity insurance is required from the sub-contractor or consultant, the amount should be stated in the right-hand column. The type of cover will be the same as stated in the contract particulars for clause 6.11. The maximum liability in part (B) will apply if any is stated. If any period other than 7 days is stated in part (D), it will also apply to clause 6.3 of the sub-contractor warranty SCWa/F.

Attestation

Alternative clauses are provided to enable the contract to be executed under hand or as a deed. The most important difference between the two is that the Limitation Act 1980 sets out a limitation period, which is ordinarily 6 years for contracts under hand and 12 years where the contract is executed as a deed. The limitation period starts to run from the date at which the breach of contract occurred. For practical purposes, the latest date from which the period would run would be the date of practical completion, this being the latest date at which the contractor could correct any breach before offering the building as completed in accordance with the contract documents: *Borough Council of South Tyneside* v. *John Mowlem & Co, Stent Foundations Ltd and Solocompact SA* (1997); *Tameside Metropolitan Borough Council* v. *Barlows Securities Group Services Ltd* (2001). Contractors will doubtless opt for contracts under hand, but employers will look to extend the contractor's liability for as long a period as possible and consequently are well advised to see that the contract is entered into as a deed.

Before the Law of Property (Miscellaneous Provisions) Act 1989 and the Companies Act 1989 came into force, it used to be necessary to seal a document in order to make it into a deed. (In Northern Ireland the need for a seal in the case of an individual was removed by The Law Reform (Miscellaneous Provisions) (Northern Ireland) Order 2005 and, in the case of a company, by the Companies (No.2) Order (Northern Ireland) 1990.) Although sealing is still possible, it is no longer necessary nor will it alone create a deed; all that is required in the case of a company is that the document must state on its face that it is a deed and it must be signed by two directors or a director and a company secretary. There are slightly different requirements in the case of an individual.

2.3 Employer's Requirements

These are the employer's instructions to the contractor. It is the information the contractor uses to prepare its proposals and if the Employer's Requirements are wrong, the Contractor's Proposals will be wrong. Essentially, this document is a performance specification. It should specify the criteria, whereas the traditional operational specification specifies the particular way in which criteria are to be satisfied. Thus, the document may specify a particular thermal insulation value, durability, load-bearing capacity and weather tightness for a wall, which the contractor can satisfy by using a number of different materials and combinations of materials. Traditionally, the actual materials and workmanship of the wall would have been specified. Although there is provision for the employer to include bills of quantities in the Employer's Requirements (supplemental provision para. 3), an employer who includes bills of quantities will throw away many of the advantages offered by the design and build concept.

It is sometimes thought that design and build is a soft option for the employer. If all that is required is a very simple building – a few thousand square metres of warehousing – the Employer's Requirements can be quite brief. In most cases, however, as much effort must be devoted to producing the performance specification as would be required for the traditional specification. The contractor is not responsible for the whole of the design but only for its completion (see Chapter 3). Therefore, the less information the employer provides, the greater will be the contractor's liability. Thus, if part of the Employer's Requirements consists of a set of very advanced working drawings, the contractor will need to do little but construct the building from those drawings and the employer will know exactly what is to be provided.

On the other hand, if the Employer's Requirements are very brief and the drawings are very simple sketch drawings, the employer will have little control over the end product. Put another way, the more that is left to the contractor, the greater will be its chance to save money and put forward an attractive tender figure. In practice, the employer will specify criteria together with any particular parts of the design which are mandatory upon the contractor, for example, marble in the lobby of a large hotel, and will make clear which aspects are left to the contractor's initiative. Contractors must beware deceptively simple Requirements which carry substantial design responsibility. In *Skanska Construction UK Ltd* v. *Egger (Barony) Ltd* (2002), the court held that the contractor was not entitled to additional payment for the supporting steelwork to a process plant, because there was sufficient, although badly defined, indication of the support steelwork on the tender drawings. This is perhaps a surprising conclusion in view of the fact that the contractor had not been supplied with loading at tender stage and, therefore, was unable to properly estimate the steelwork required.

It is very important that the employer crystallises the Requirements before executing the contract. Although provision is made in the contract for the employer to make changes in the Requirements, it is by no means as easy to do this as it is in a contract such as SBC, and the contractor will have the right to object to many changes (see Chapter 10). An employer who might wish to make changes once the construction has begun should seriously consider using another more suitable form of contract, because apart from other considerations, the employer will lose many of the advantages, in terms of risk and price, offered by this form (see Chapter 1).

It is strongly advised that the employer obtains planning permission before accepting any tender. It is perfectly possible to make the contractor responsible for obtaining such permission, but it should be remembered that actually getting permission can never be guaranteed, because it depends upon the planning authority. Therefore, the situation could arise where the contractor applies unsuccessfully for planning permission, or if successful, it may take several months of negotiation before it is finalised. Not only does the contract make provision for extension of time in such cases (clause 2.26.12) – that is, after all only reasonable – it also entitles the contractor to loss and expense (clause 4.20.4), which is also reasonable. It is possible for the employer to specify that amendments to comply with planning requirements are not to be treated as changes in the Employer's Requirements and, therefore, are to be carried out at the contractor's own cost, but probably there will be a hefty price to pay at tender stage.

There are two points which merit careful attention. The first point is that many of the statements made by the employer within the Employer's Requirements will be representations. The contractor will use the information in compiling its tender. Typically, this will include information about the site and ground conditions. If any of the statements of fact are incorrect, they will probably amount to misrepresentations. A misrepresentation which is one of the inducing causes of a contract and which causes loss to the innocent party may result in legal liability.

Depending on whether the misrepresentation is innocent, negligent or fraudulent, the contractor may be able to recover damages or even to put the contract at an end if it suffers some loss thereby. The employer may not necessarily be able to avoid the consequences of a misrepresentation by including a warning to the contractor to check, or even by including a disclaimer. It may still be held to be a misrepresentation for which the contractor has a remedy in law: *Cremdean Properties Ltd and Another* v. *Nash and Others* (1977). It is difficult for the employer to avoid liability for statements in the Employer's Requirements and any attempt to do so should be drafted only after receiving proper advice.

A misrepresentation may also amount to a collateral warranty. For example, in *Bacal Construction (Midland) Ltd* v. *Northampton Development Corporation* (1976), which involved a design and build contract, the contractor was instructed to design foundations on the basis that the soil conditions were as indicated in borehole data provided by the employer. The Court of Appeal held that there was a collateral warranty that the ground conditions would be in accordance with the hypotheses upon which Bacal had been instructed to design the foundations, and held that they were entitled to damages for its breach.

The second point is that many sets of Employer's Requirements contain a provision to the effect that workmanship and/or materials are to be to the employer's approval. The result of inserting such a provision is that when the final account and final statement become conclusive as to the balance due between employer and contractor, they are also conclusive evidence that any materials or workmanship reserved for the employer's approval are to the employer's reasonable satisfaction, subject to very limited exceptions (clause 1.9.1.1). This provision makes it difficult for the employer to contend subsequently that such materials or workmanship are defective. It should be noted that it does not matter whether the employer has, in fact, actively taken steps to be satisfied about the materials or workmanship. If nothing is reserved to the employer's approval, the pitfall is avoided. Other phrases such as 'to the employer's satisfaction' may well have the same effect. It is recognised that there will be situations in which the employer will insist on reserving final approval rather than relying on any performance criteria. Such situations should be limited and the employer or the employer's agent must make sure that the items in question are carefully inspected before practical completion and again before the final account and final statement become conclusive. The particular wording of the contract appears to avoid the highlighted situation in IFC 84 and JCT 80 where the final certificate was conclusive regarding the architect's opinion of quality and standards whether or not expressly reserved to the architect's opinion: *Colbart* v. *Kumar* (1992); *Crown Estates Commissioners* v. *John Mowlem & Co* (1994). In any event, in an excess of caution, JCT probably settled the matter by the issue of Amendment 9 in 1995. The following matters should always be included in the Employer's Requirements:

- Details of the site including the boundaries (unless the site is being provided by the contractor, in which case clause 2.9 must be amended).
- Details of accommodation requirements.
- Purposes for which the building is to be used.
- Any other matter likely to affect the preparation of the Contractor's Proposals or its price.
- Statement of functional and ancillary requirements

  — Kind and number of buildings
  — Density and mix of dwellings and any height limitations
  — Schematic layout and/or drawings
  — Specific requirements as to finishes etc.

- Bills of quantities in accordance with supplemental provision para. 3, if required.
- Details of any provisional sums.
- Statement of planning and other constraints, e.g. restrictive covenants, together with copies of any statutory or other permissions relating to the development.
- Statement of site requirements.
- The extent to which the contractor is to base its proposals on information supplied in the Employer's Requirements.
- Access restrictions.
- Availability of public utilities.
- Details of the contractor's programme required.
- The method of presentation of the Contractor's Proposals

  — Drawings, plans, sections, elevations, details, scales
  — Any special requirements, for example, models, computer animation, video
  — Layout of specialist systems
  — Specification requirements.

- If schedule 2 is not to be used, the Employer's requirements regarding submission of contractor's drawings.
- If supplemental provision para. 1 is used, the employer's requirements regarding the records the site manager is required to keep.
- Detailed requirements in respect of the as-built drawings which the contractor must supply in accordance with clause 2.37.
- Whether stage or periodic payments are to be made.
- Functions to be carried out by the employer's agent and, if required, the quantity surveyor and the clerk of works.
- Information to be included for the contract:

  — The form of the Contract Sum Analysis and its content
  — Whether the employer is a 'contractor' under the CIS
  — The name of the adjudicator and the nominator
  — If arbitration or litigation is to apply and if arbitration, the appointor of the arbitrator
  — The method of fixing the date for completion
  — The base date

— If dwellings, whether subject to the NHBC scheme
— Whether and to what extent there is any limit on the contractor's liability for consequential loss
— The detailed manner in which the contract particulars are to be completed
— List of materials for clause 4.15 and whether a bond is required
— Whether advanced payment will be made and whether a bond is required.

Supplemental provision para. 3 sets out certain rules if the Works are described in the Employer's Requirements by bills of quantities:

- The method of measurement must be stated.
- Errors in the bills must be corrected by the employer and the correction is to be treated as if it were a change in the Employer's Requirements.
- If a valuation is carried out under the terms of clause 5.4 to 5.7, rates and prices in the bills of quantities must be substituted for the reference to values in the Contract Sum Analysis.
- If price adjustment formulae are to be used (schedule 7), the rates and prices in the bills of quantities are to be used so far as is relevant.

For the employer to include bills of quantities in the Employer's Requirements indicates that the design of the building is very advanced. If that is the case, the amount of design left to the contractor will be very small. In that situation, it may well be advisable for the employer to continue with the project on a traditional basis. The employer could reasonably ask the contractor to provide bills of quantities as part of the Contract Sum Analysis, but they seem to have no logical place in the Employer's Requirements.

2.4 *Contractor's Proposals*

Put simply, the Contractor's Proposals should answer the Employer's Requirements. If the Employer's Requirements are detailed, the Proposals will be similarly detailed. If the Requirements are rather vague, the Proposals may well leave many loose ends and there are likely to be elements of the building which are not quite what the employer expected. Therefore, to take an extreme case, if the employer simply asked to be provided with 30,000 square metres of office space on a particular site, it will leave the contractor with tremendous scope in design, construction and costing.

Most contractors will submit a detailed specification covering all the work and materials they will use to complete the project. They may also include a programme and a method statement. It is not usual to make either of these documents a contract document, because to do so requires both employer and contractor to comply with it in every particular. If it becomes necessary for the contractor to carry out the work in a different way, it may be entitled to claim payment: *Yorkshire Water Authority* v. *Sir Alfred McAlpine & Son (Northern) Ltd* (1985).

The contractor must plug any gaps in the Employer's Requirements by including the information in its Proposals. This is particularly true about the contract data. If an important point such as the system of payment has been omitted from both documents, there is a ready-made source of dispute before the contract is executed.

It sometimes happens that the contractor wishes to propose a material or constructional detail which is contrary to what is contained in the Employer's Requirements. The two documents must be consistent and, therefore, the contractor must draw the employer's attention specifically to such a proposal so that, if accepted, the Requirements can be amended before the contract documents are signed. The contractor is best advised to make such a proposal as an alternative and subject to a stated price adjustment.

The Contractor's Proposals should not contain any provisional sums unless they are in the Employer's Requirements. If the contractor feels that a provisional sum must be included, although not requested by the employer, the employer's attention again must be drawn to the sum so that it can be included in the Employer's Requirements document before signing. The consequences of discrepancies are discussed in section 2.7.

2.5 *Contract Sum Analysis*

DB is a lump sum contract. This means that, essentially, the contractor carries out the work for a fixed and stated amount of money payable by the employer. There is no provision for re-measurement, although there is provision for changes in the Employer's Requirements and fluctuations. Payment may be made by fixed stages or by periodic payment based on the value of work done. The purpose of the Contract Sum Analysis is to assist in valuation of changes and work carried out, where appropriate, and to enable fluctuations to be calculated. The employer may require the Analysis in any form and the contractor must comply. Where formula fluctuations are to be used, the Analysis must contain the appropriate information, properly arranged. Whether or not the employer so requires, the contractor should always include a method of valuing design work. This might very likely be on an hourly basis and it will be needed in the valuation of changes and also in the valuation of design work carried out but later aborted. This is a common occurrence in design and build where the employer may ask the contractor to suggest alternative designs for part of the building, but eventually may proceed with the original design on which the contractor's price was based. In the absence of a clearly laid down system of charging for such work, the contractor may find that it recovers nothing or, at best, a nominal amount.

2.6 *Supplementary provisions*

The supplementary provisions were issued originally as part of amendment 3 to JCT 81 in February 1988. They were proposed by the British Property Federation and there are marked similarities between the provisions and certain clauses in the BPF's own form of contract. The provisions have changed somewhat since they were first introduced and are now as follows:

1. Site manager
2. Persons named as sub-contractors in the Employer's Requirements
3. Bills of quantities

4. Valuation of change instructions – direct loss and/or expense – submission of estimates by the contractor
5. Direct loss and/or expense – submission of estimates by contractor.

The provisions will be dealt with throughout the book under the various topic headings as appropriate.

The supplemental provisions are contained in schedule 2 and, if the employer wishes them to apply, the appropriate part of the contract particulars must be completed. It is suggested that the employer would be wise to complete the contract particulars so that the provisions do apply, because they are generally very sensible. If some of the provisions are not required, they should be deleted. Contractors must be wary to see where they do apply and, if so, note every significant effect on the contract.

2.7 *Priority, discrepancies, errors*

Clause 1.3 provides that the contract is to be read as a whole and that nothing in the Employer's Requirements, the Contractor's Proposals or the Contract Sum Analysis overrides or modifies the articles of agreement or the conditions. The effect in practice is that if there is a conflict between a term in the printed contract and a term in the Employer's Requirements, say differing periods of notice under clause 3.6, the period in the printed form will apply. This type of clause has been upheld in the courts: see, for example, *M J Gleeson (Contractors) Ltd* v. *Hillingdon Borough Council* (1970); *English Industrial Estates Corporation* v. *George Wimpey & Co Ltd* (1973). In the absence of this clause, the ordinary rule of interpretation would apply, namely that where a contract is contained in a printed form and there is inconsistency between the printed terms and typewritten terms, the typewritten terms would prevail. That sensibly assumes that if the parties have a set of contract documents consisting of a standard printed form and a typed or written section, the typed or written section would prevail in the case of any conflict. In order to amend a printed clause it is necessary to amend it on the form itself and have both parties initial the amendment. Another way is to have any special clauses initialled by the parties and annexed to the printed form, with an appropriate reference inserted in the articles.

The simplest way of removing the problem is to delete the second part of clause 1.3 (after 'whole'). Care must be taken not to fall into the trap of simply stating in the Employer's Requirements that the relevant part of the clause is deleted, without actually deleting it in the form! It is very common and correct for any amendments to the contract clauses to be listed in the Employer's Requirements. Where this occurs, and if the relevant part of clause 1.3 is not deleted, the employer's professional advisor must ensure that these amendments are meticulously transferred to the printed form before the contract is executed.

Clauses 2.10, 2.12, 2.13 and 2.14 deal with discrepancies. Clause 2.9 provides that the employer must define the site boundaries. Clause 2.10.1 reasonably provides that if there is a divergence between what the employer has defined and anything contained in the Requirements, the employer must issue an instruction to correct the matter, which is deemed to be a change. If either the employer or the contractor finds the divergence, one must give the other a written notice (clause 2.10.2). Clause 2.13 provides that if the contractor finds any inadequacy, discrepancy or divergence in or between the Employer's Requirements, Contractor's Proposals, an instruction of the employer (other than for a change) or information issued by the contractor under clause 2.8, it must immediately give written notice to the employer and the employer must give instructions. It is now established that the contractor has no duty to look for or to find such divergences in this or other instances, but simply to give notice if it finds them: *London Borough of Merton* v. *Stanley Hugh Leach Ltd* (1985). If the contractor is itself to provide the site, clause 2.10.1 must be amended. It is suggested that the amended clause should make reference to the Contractor's Proposals, instead of the Employer's Requirements, and to the definition of site boundary given by the contractor.

Clause 2.14.1 deals with discrepancies within the Contractor's Proposals. The contractor must immediately inform the employer in writing giving details of its suggested amendment. The employer may then choose between the discrepant items or it may choose the contractor's suggestion, all at no additional cost. The employer must take care to give a decision within a reasonable time or the contractor will have grounds for extension of time and loss and/or expense; this situation is covered in clauses 2.26.5 and 4.20.5 respectively. An employer who dislikes all the available options may issue a change instruction under clause 3.9, but that is not a prudent course unless absolutely necessary because the employer pays a premium in the cost of the change and, possibly, extension of time and loss and/or expense. The results of failure by either party to note the discrepancy before construction would be firmly at the cost of the contractor.

Clause 2.14.2 deals with the position if there is a discrepancy within the Employer's Requirements or between the Requirements and any change issued in accordance with clause 3.9. The reference to the change is intended to cover the situation where the employer issues a change instruction which, while obviously changing the particular part of the Requirements at which it is aimed, inadvertently conflicts with something else which is not the subject of the change. In the case of any such discrepancies, if the matter is addressed within the Contractor's Proposals then they will prevail and there will be no additional costs to the employer, neither will there be any reduction even if the treatment in the Contractor's Proposals is clearly less expensive than either of the discrepant items.

For example, both walnut panelling and plastic-faced steel panelling may be separately required for the boardroom. If the Contractor's Proposals allow for only plaster, the Contractor's Proposals for plaster prevail. That does not mean that the employer is obliged to have plaster on the boardroom walls, but that if either of the more expensive finishes is required, the employer will have to pay for it. If the Contractor's Proposals do not deal with the matter, the contractor is required to give the employer written notification of its amendment to resolve the discrepancy. The employer may either accept the amendment or decide on a different solution. In either case, the employer's decision is to be treated as a change which will be valued under clause 5. In addition, the contractor may be entitled to an extension of time under clause 2.26.1 and direct loss and/or expense under clause 4.20.1. If the decision by the employer is late and causes delay or disruption to the contractor, that again is grounds for both extension of time and loss and/or expense. It is thought that, if the discrepancy was not detected until the element was constructed, the employer must issue a change instruction to correct the problem and the contractor would be entitled to reimbursement in terms of money, time and loss and/or expense.

What is the situation if there is a discrepancy between the Employer's Requirements and the Contractor's Proposals? Unfortunately, the contract still does not expressly give an answer to that question. That is a serious omission and one that has been noted in earlier editions of this book. It is often suggested that the third recital comprehensively deals with the matter in favour of the Contractor's Proposals, because it states that the employer has examined the Contractor's Proposals and the Contract Sum Analysis and, subject to the conditions, the employer is satisfied that they appear to meet the requirements. Such arguments seem to be seriously flawed. The recitals cannot be used to modify clear words in the body of the contract. It is only when the operative part of the contract is ambiguous that the recitals can be called on to help determine the correct meaning: *Rutter* v. *Charles Sharpe & Co Ltd* (1979). It will be seen below that the contract is quite clear about the priority, albeit it takes a little digging among the clauses to establish the position.

There are three important points to note about the third recital. The first is the use of the word 'appear'. This makes clear that the satisfaction is in looking at the general impression. It is obviously not intended that the employer is expected comprehensively to check the Contractor's Proposals to make sure that they meet the Employer's Requirements. Had that been intended, clear words could have been used to that effect. The third recital merely records that the employer believes that the Contractor's Proposals provide what was requested in the Employer's Requirements; put bluntly, that if the employer asked for a concert hall, that is what the Proposal is about; that if the employer was adamant about having a steeply pitched roof, the Proposals provide exactly that. It cannot be expected that the employer will count the numbers of coat hangers or undertake a thorough comparison. Indeed, the former practice note issued for JCT 81 made that clear, although of course the practice note was not part of the contract.

The second point to note is that the third recital is expressly subject to the conditions. The fact that the statement is made subject to the conditions informs the user that the printed conditions have something important to say. What the conditions say will be examined below.

The third point to note about the third recital is footnote [3]. It says that if the employer has accepted a divergence from the Employer's Requirements in the Contractor's Proposals, the divergence should be dealt with in the Employer's Requirements before the documents are signed. If the divergence still exists after signing it suggests, to put it at its weakest, that the employer has not accepted it, probably because it has not been noticed. This merely confirms the philosophy of the contract as can be discerned from the recitals as a whole. The first recital states that the employer has supplied the Employer's Requirements to the contractor and the second recital states that the Contractor's Proposals have been supplied to the employer in response. Therefore, it is clear that the Contractor's Proposals should show how the contractor is to comply with the Employer's Requirements.

As it stands, the third recital is intended to indicate that the employer accepts that, at face value, the Contractor's Proposals respond to the stated criteria, but the employer is reserving the position as regards the actual satisfaction of such criteria. There is something to be said for this approach, because the employer, whether or not professionally advised, cannot be expected to carry out detailed checks of the

Proposals. Any 'approvals' given by the employer must be seen in this light: *Hampshire County Council* v. *Stanley Hugh Leach Ltd* (1991).

The limited use of recitals has already been pointed out. In fact, there is no need to refer to the recitals, because the operative part of the contract, mainly the conditions, puts the matter beyond any reasonable doubt. The wording of the contract strongly points to the intention that the Contractor's Proposals will be drafted to meet the Employer's Requirements. Consideration of the following points leaves no room for doubt that the Employer's Requirements prevail:

- Clause 2.2 provides that all materials and goods for the Works are to be of the kinds and standards described in the Employer's Requirements; it is only if the Employer's Requirements make no mention of the materials and goods, that the contractor can refer to the specification in the Contractor's Proposals. Clause 2.2.2 deals with workmanship in virtually identical terms.
- Importantly, under the terms of the contract, the employer cannot issue a change instructing the contractor to vary the Contractor's Proposals. Clause 5.1 unequivocally provides that a change means a change in the Employer's Requirements. Moreover, under clause 5.2.3, the employer is not entitled to instruct the expenditure of a provisional sum in the Contractor's Proposals. If the Contractor's Proposals prevailed over the Employer's Requirements, the employer would be unable to issue changes to deal with any part of those Contractor's Proposals. For example, if the Employer's Requirements called for bench seating in a waiting area and the contractor proposed separate chairs, the employer would be unable to restore the seating to benches by issuing a change, because it would not be a change in the Requirements. The Requirements would be the same as always. If the Contractor's Proposals were to take precedence, the employer would have to be able to issue a change instruction to change the Contractor's Proposals; in the case of this example, from chairs to benches. Because the Employer's Requirements take precedence, there is no need to issue a change instruction at all; the contractor is obliged to comply with the Employer's Requirements. All that is needed is a note to the contractor, pointing out the error. As soon as instructions regarding order of work, working space and access are taken into account, the primacy of the Employer's Requirements is beyond doubt.

Obviously, the Employer's Requirements and the Contractor's Proposals should read as one. The simplest way to tackle the problem is to insert a clause to the effect that if there is any discrepancy between the Employer's Requirements and the Contractor's Proposals, the Employer's Requirements will take precedence.

If the contractor makes a unilateral error in its Proposals or in the Contract Sum Analysis, e.g. errors in pricing, it will have to stand the consequences unless the employer or the professional advisors discover the error before acceptance and realise that it is not intentional: *W Higgins Ltd* v. *Northampton Corporation* [1927]. This may be thought a harsh view, but the contractor undertakes a very great burden of responsibility under this form of contract. After all, that is the attraction so far as the employer is concerned. The contractor may possibly get some relief if it can demonstrate that the employer knew of the error at the time the tender was accepted: *McMaster University* v. *Wilchar Construction Ltd* (1971), a decision of the Ontario High Court.

2.8 Custody and copies

Clause 2.7.2 provides that immediately after the execution of the contract, the employer must provide the contractor with one copy of each of the contract documents. Each document must be certified on behalf of the employer (therefore, presumably by the employer's agent). It is quite sufficient for the employer's agent to write on each document: 'Certified a true copy of the . . .' and to sign and date it. Custody of the contract documents is to be the responsibility of the employer in accordance with clause 27.1.1. Sometimes the employer may fail to provide the certified copies and employers have been known to refuse to provide them even when requested. The prudent contractor will always keep a copy of the documents it signs against just such an eventuality. The employer's failure would no doubt be susceptible to a mandatory injunction if the contractor had no copy and the employer persistently refused to supply one, but in the event of a dispute which went to arbitration or litigation, the documents would be subject to discovery.

The contractor has a duty under clause 2.8 to supply the employer with two copies of all the drawings, specifications and other information which it either prepares or uses for the purposes of the Works (referred to as the 'Contractor's Design Documents'). Thus, strictly, the contractor must supply information prepared for the works even if ultimately unused. The information must be submitted in accordance with the design submission procedure set out in schedule 1 or any other procedure which is otherwise stated in the contract documents (see chapter 4.4.1) The employer's own procedure will not fall foul of the priority clause 1.3, because clause 2.8 expressly makes provision for it. The contractor must keep one copy of all this information, together with the Employer's Requirements, the Contractor's Proposals and the Contract Sum Analysis, on site so that the employer's agent can have access to them at all reasonable times (clause 2.7.3).

Clause 2.37 is important and it is vital to understand exactly what it means. It stipulates that before practical completion of the Works or any section, the contractor must supply the employer with whatever drawings and information are specified in the Employer's Requirements and the Contractor's Proposals or as the employer may reasonably require relating to the Works as built. The provision is expressly stated not to affect the contractor's liabilities under clause 3.18 in relation to the CDM Regulations, and the health and safety file with regard to the maintenance and operation of the Works including any installations (presumably such things as heating systems). The important point is that the contractor's obligation is essentially to supply what are commonly known, and indeed stated in the heading, as 'as-built' drawings. The usual prohibition against divulging the contents of contract documents and other confidential information is put on both parties by clause 2.7.4, with the exception of any information which the employer wishes to use in connection with maintenance, use, repair, advertisement and letting or sale of the Works.

2.9 Notices

Clause 1.7 of the contract sets out the requirements for the giving or service of notices or other documents. It applies if the contract does not expressly state the way in which service of documents is to be achieved. Therefore, it does not apply to notices given in connection with the termination procedures in clause 8, because those clauses state that service is to be carried out by means of actual delivery, special or recorded delivery. In other cases, service is to be by any effective means to the address in the contract particulars or to any agreed address. Surprisingly, parties are quite capable of squabbling over service and appropriate addresses. If that is the situation, service can be achieved by addressing the document to the last known principal business address, or if the addressee is a body corporate, to that body's registered office or its principal office, provided it is prepaid and sent by post.

The contract also provides, in clause 1.8, for electronic communication if the parties so wish. If they do so wish, they can insert appropriate details into the contract particulars. This is a very sensible option and it is in line with current practice. However, it should be noted that it is not permissible to send all communications electronically and some confusion is possible. Clause 1.8 is subject to 'the specific provisions referred to in clause 1.7'. Clause 1.7 itself is made subject to the specific provisions of the contract about the manner of giving any notice or document. What it amounts to is that if the parties wish to communicate electronically, they may do so with two provisos:

- Communications may not be made electronically if the contract specifically states they are to be made in some other way; and
- Details of the communications which the parties wish to send electronically must be listed in the contract particulars.

The confusion arises in this way: electronic communication seems to be a 'manner' of serving, just as service by actual delivery or by post is a 'manner' of service mentioned under clause 1.7. Few would deny that a document sent by post or given by actual delivery is in writing. Indeed, a document sent by fax has been held to be in writing and constituting 'actual delivery': *Construction Partnership UK Ltd* v. *Leek Developments Ltd* (2006). Notice served by e-mail was held to be in writing and valid for the purposes of the Arbitration Act 1996: *Bernuth Lines Ltd* v. *High Seas Shipping Ltd* (2006). Therefore, it appears that all communications might be made electronically other than clause 8 notices which must be given by actual, special or recorded delivery. However, reference to the contract particulars apparently restricts electronic communication to communications which the contract does not require to be 'in writing'. That would preclude, among other things, all instructions, withholding notices and clause 2.24.1 delay notices and written statements of partial possession, but curiously and inconsistently, it would allow the practical completion statement to be served electronically. Some slight re-drafting would be appropriate to resolve the conflict between the contract particulars and clauses 1.7 and 1.8, and in light of recent judgments. (See also section 12.2 regarding actual delivery.) It is probable that an application for payment could be validly issued by e-mail if it was clear that the parties were communicating, or were expecting to communicate, in this way even if the electronic communications section of the contract particulars was not completed: *Palmac Contracting Ltd* v. *Park Lane Estates Ltd* (2005).

Clause 1.5 usefully sets out the way in which periods of days are to be reckoned. This is to comply with the Housing Grants, Construction and Regeneration Act 1996. If something must be done within a certain number of days from a particular date, the period begins on the day after that date. Days which are public holidays are excluded. Clause 1.1 helpfully defines public holidays as 'Christmas Day, Good Friday or a day which under the Banking and Financial Dealings Act 1971 is a bank holiday'. A footnote instructs the user to amend the definition if different public holidays apply.

Clause 1.11 states that the law applicable to the contract is to be the law of England. That will be the case no matter that the nationality, residence or domicile of any of the parties is elsewhere. Where a different system of law is required, this clause must be amended. For example, if the Works are being carried out in Northern Ireland, the parties will probably wish the applicable law to be the law of Northern Ireland. Curiously, and despite this inconsistency being highlighted in previous editions, the applicable law of the two bonds which are bound into the contract in schedule 6 is still stated to be the law of England and Wales.

Chapter 3
Design Liability

3.1 *General principles of design liability*

3.1.1 Basic principles of liability

The liabilities of a designer are in principle no different from the liabilities of any person, i.e. the designer may have liabilities in contract or in tort.

The general principle of contract is that the parties to the contract have agreed on mutual rights and duties which they would not otherwise have in law. The parties may agree any terms they wish provided only that they are lawful. The terms of a contract between an architect and the client will usually include a term that the architect will use reasonable skill and care in the execution of duties under the contract. Such a term will usually be express, but it may also be implied since anyone holding him or herself out as an architect impliedly warrants the possession of the necessary ability and skill. Terms may be implied into a construction contract, or indeed into a contract for professional services, either as a matter of law or as a matter of fact (see section 4.2).

The classic modern statement on the doctrine of implication of terms is that of Lord Simon in *BP Refinery Ltd* v. *Shire of Hastings* (1978), where he said:

> '[For] a term to be implied, the following conditions (which may overlap) must be satisfied: (1) It must be reasonable and equitable; (2) it must be necessary to give business efficacy to the contract, so no terms will be implied if the contract is effective without it; (3) it must be so obvious that "it goes without saying"; (4) it must be capable of clear expression; (5) it must not contradict any express term of the contract.'

There are limits to when terms will be implied. For example, a term will not be implied at common law simply because the court thinks it is reasonable to insert it into the contract, and even where one or other of the situations referred to above may otherwise arise, terms will generally be implied only under certain conditions. Therefore, an implied term must not be in conflict with, or inconsistent with, an express term and it must be based on the imputed or presumed intention of the parties: *Liverpool City Council* v. *Irwin* (1977).

An architect's duties may be quite extensive and include far more than just design (see, for example the RIBA Standard Conditions for the Appointment of an Architect (CA-S-07-A)). In the case of design and build, typical design contracts would be between an architect and the client in the early stages of a project, then in the later stages of a project between the contractor and the employer and possibly between the contractor and an architect or other construction professional. Any formal collateral warranty entered into is also a contract (see section 3.7).

A failure by a party to a contract to carry out its duties regarding design would be a breach of contract which would entitle the other party to recover damages in respect of any proven loss, and it matters not whether the designer is an architect or other design professional or a design and build contractor. The principle of recovery is that the injured party should be put in the position, so far as money can, as if the breach had not been committed (*Robinson* v. *Harman* (1848)) although this may be modified in practice: *Ruxley Electronics & Construction Ltd* v. *Forsyth* (1995). In contract, there are two kinds of damages which can be recovered:

- Damages that may fairly and reasonably be considered to arise naturally out of the breach; and
- Damages which are the result of special circumstances known to the parties at the time the contract was entered into and which are capable of causing a greater loss than otherwise would be the case: *Hadley* v. *Baxendale* (1854).

A designer whose design fails may also be liable in tort for negligence. In light of recent developments in the law, it is thought that where there is a contract between the parties, there will be a parallel liability owed in tort to the other party for any resulting loss which may be wider in scope than the liability in the contract: *Holt* v. *Payne Skillington* (1995). This is so whether the negligent designer is an architect or other design professional or a design and build contractor.

Economic loss is distinct from damage which results from physical injury to, or death of, a person or physical damage to property other than the building itself.

It used to be common for a building owner whose building failed, to bring actions in both contract and in tort, the one claiming damages for breach of contract and the other claiming damages for negligence. It is sometimes difficult to decide whether a particular loss is to be categorised as economic and the courts have always tended to make the decision on the grounds of policy as much as anything else, because in the last analysis, it is possible to say that most loss is economic: *Spartan Steel & Alloys Ltd* v. *Martin & Co (Contractors) Ltd* (1972). A significant change came about in *Murphy* v. *Brentwood District Council* (1990) and, as a result of that case, it has been well said that it is now 'necessary to read all English authorities concerning negligence decided between 1971 and 1990 with extreme caution': *Keating on Building Contracts*, 7th edn, 2001, 7.01.

However, although it is now clear that normally a plaintiff cannot recover economic loss in an action for negligence, but must establish actual death or physical injury to persons or property other than the defective building itself, economic loss can be recovered.

A 'reliance situation' will seemingly only arise if there is a special relationship of both proximity and reliance between the parties. Where the representor has some special skill or knowledge and gives advice about it, and the representor knows, or it is reasonably foreseeable, that the other will rely on the advice and the advice has been acted on by the other, the resultant economic loss will be recoverable provided that it is foreseeable: *Hedley Byrne & Co* v. *Heller & Partners Ltd* (1963).

The House of Lords have considered and restricted the criteria for the special relationship. The current position appears to be that a special relationship will be considered to exist if:

- The advice is required for a purpose which is made known to the misrepresentor at the time the advice is given; and
- The misrepresentor knows or can reasonably foresee that the advice will be communicated to the other either personally or as a member of an ascertainable class in order to be used for the purpose initially made known; and
- It is known by the misrepresentor that the advice is likely to be acted upon by the other without further enquiry; and
- It is so acted upon and the other suffers some detriment: *Caparo Industries plc* v. *Dickman and Others* (1990).

Because special skills and advice are features of this kind of liability, it is often associated with professional advice, although not necessarily: *Barclays Bank* v. *Fairclough Building Ltd and Carne (Structural Repairs) Co Ltd and Trendleway Ltd* (1995).

At one time, the case of *Junior Books Ltd* v. *The Veitchi Co Ltd* (1982) appeared to open the floodgates to the recovery of economic loss, but in subsequent cases it has been distinguished almost to the point of extinction: *Muirhead* v. *Industrial Tank Specialities Ltd* (1986). However, it has not been directly overruled and, indeed, in *Murphy* v. *Brentwood District Council* it was explained by some members of the House of Lords as resting upon the *Hedley Byrne* principle of reliance, although on what basis that 'reliance' arose is not easy to see. The House of Lords appeared to treat the employer/nominated sub-contractor relationship as an almost unique situation

There was no collateral contract involved in the *Junior Books* case, which involved a nominated sub-contractor – and the relationship between a nominated sub-contractor and an employer is hardly, on any reasonable view, a 'special relationship'. It is thought that the decision is not of general application and would certainly not be extended.

The current position, as determined by the House of Lords in the important case of *Murphy* v. *Brentwood District Council* (1990), may be broadly summarised, in the context of the construction industry, as follows:

- Negligence which results in a defect in the building itself is not actionable in tort. There is no actionable damage to 'other property'.
- To be actionable, the defective structure must cause damage to other property or result in death or personal injury.
- If, however, the defect is discovered before it has caused damage, the cost of making good the defect is not recoverable.
- In a complex structure such as a building, it is not permitted to consider the building as a series of segments, one causing damage to another – the so-called 'complex structure' theory.
- However, there is possibly potential liability where 'some distinct item incorporated in the structure . . . positively malfunctions so as to inflict positive damage on the structure in which it is incorporated'. For example, where a sub-contractor installs a central heating boiler into a building and it explodes, causing damage to the building, the negligent installer might be held liable in damages, or where subsequent underpinning is the cause of damage to the original building: *Jacobs* v. *Morton & Partners* (1994).

The reasoning which led to the *Murphy* decision was given earlier effect in *Pacific Associates Inc* v. *Baxter* (1988), where contractors were unsuccessful in an action for negligence against a supervising engineer, under a FIDIC contract, who had failed to certify certain sums allegedly due to them. The contractors also alleged unsuccessfully a breach by the engineers of a duty to act impartially.

The Court of Appeal, in dismissing the contractors' claim, stressed the importance of the terms of the contract between the employer and the contractor, which provided for arbitration of disputes arising under the contract. In stressing the importance of the contractual route, the following principles appear to have been important:

- The engineers were agents of the employer.
- The engineers had a contractual obligation to the employer to use skill and care and to act fairly between the parties.
- The contractors had relied on their remedies against the employer under the contract between them by going to arbitration over the disputed claims. In the event, the arbitration was settled on terms of an *ex gratia* payment by the employer to the contractors who were attempting to recover the shortfall between the amount claimed and the amount of the settlement by means of an action in tort against the engineers.
- The engineers had not assumed responsibility to the contractors for economic loss resulting from a breach of any of the obligations in the contract between the employer and the contractors.
- Therefore, there was no basis, either on the *Hedley Byrne* principle or otherwise, by which the engineers could be said to owe a duty of care to the contractors.
- There was an arbitration provision which enabled the contractor to recover from the employer. The engineers had a duty to act in accordance with the construction contract, but that duty arose from the contract between the engineers and the employer. The contractors could challenge the performance of the engineers by claiming against the employer. There was, therefore, no justification for imposing on the contractual structure an additional liability in tort.
- There was an exclusion clause in the main contract, although not something to which the engineers were a party.

This appears to be of great importance to architects and others advising either employer or contractor and the principles were also applied by the High Court of Hong Kong to an architect under a building contract. However, doubt has been thrown on this decision (see the discussion in *Parris's Standard Form of Building Contract* by David Chappell, 3rd edition (2002) Blackwell Publishing, pp. 58–61), even if, as it seemed, it depended on its own special facts. It is probably dangerous to assume that this case provides a blanket protection for all engineers and architects. It is still possible that the contractor might be able to recover against a negligent or unfair or partial architect or other certifier who deliberately or negligently under-certifies amounts due.

Subsequent developments in the way the courts have interpreted the *Hedley Byrne* principle suggest that actions against architects based on reliance may not be far away. It has been held that the principle extends beyond the provision of information to include the performance of other services (*Henderson* v. *Merrett Syndicates Ltd* (1994)) and that there is no sustainable distinction between the making of statements and other exercises of duty: *Conway* v. *Crowe Kelsey & Partner* (1994). In *J Jarvis & Sons Ltd* v. *Castle Wharf Developments* (2001), the Court of Appeal took the view that the employer's professional agent could be liable to a contractor for negligent misstatements if they were made in order to induce the contractor to tender and if the contractor relied on them.

3.1.2 Designer's position

The professional designer, such as an architect or an engineer, is required to exercise reasonable skill and care: *Lanphier* v. *Phipos* (1938). The designer is not required to guarantee the result unless also providing the end product, such as a dentist making a set of false teeth: *Samuels* v. *Davis* (1943). A designer who fails to exercise the requisite amount of skill and care may be negligent:

> 'Where you get a situation which involves the use of some special skill and competence, then the test as to whether there has been negligence or not is not the test of the man at the top of the Clapham omnibus, because he has not got this special skill. The test is the standard of the ordinary skilled man exercising and professing to have that skill; it is well established law that it is sufficient if he exercises the ordinary skill of an ordinary competent man exercising that particular art.' (*Bolam* v. *Friern Hospital Management Committee* (1957)).

Thus, a designer will be thought to act correctly if acting with the kind of skill an average designer would display. In order to decide such issues, the court must hear expert testimony from other designers on the matter in question. It is not always sufficient for a designer to be able to maintain that he or she simply did the same as other designers were doing, if it can be shown that generally accepted practice is not correct: *Sidaway* v. *Governors of the Bethlem Royal Hospital and the Maudsley Hospital* (1985).

A person will be judged on the basis of the skills that person professes to have. Therefore, if an architect purports to have special expertise in the restoration of old buildings, a failure in that respect will be compared to the performance of other architects who have that special expertise, whether or not the original architect does in fact have such expertise.

On the other hand, the standards to be applied to a professional person will be the standard of the time when the professional acted and not the standards commonly practised at some time after the act: *Wimpey Construction UK Ltd* v. *D V Poole* (1984). This is usually referred to as the 'state of the art' defence.

Moreover, the designer's responsibility does not end when the design is completed and handed to the builder. There is a continuing responsibility to review and revise the design if any problems become apparent: *London Borough of Merton* v. *Lowe & Pickford* (1981). It appears that this duty ends after the designer's initial involvement ends, so that the designer is not burdened with the responsibility of constantly reviewing designs thereafter: *T E Eckersley and Others* v. *Binnie & Partners and Others* (1988). It is arguable that the designer's duty to review the design when put on notice continues until the final certificate or other final accounting process is complete: *New Islington & Hackney Housing Association* v. *Pollard Thomas & Edwards* (2001).

The designer who uses untried methods of construction or materials will be just as liable for design failures as in the case of a failure while using well tried methods and materials. The designer cannot blame general lack of knowledge as a basis for a state of the art defence. Special care is needed, therefore, before new techniques are put into operation:

> 'For Architects to use untried or relatively untried materials or techniques cannot in itself be wrong, as otherwise the construction industry can never make any progress. I think however, that Architects who are venturing into the untried or little tried would be wise to warn their clients specifically and get their express approval.' (*Victoria University of Manchester* v. *Hugh Wilson & Lewis Womersley and Pochin (Contractors) Ltd* (1984))

3.1.3 Fitness for purpose

It has already been stated that the law will require a professional person, such as an architect or an engineer, to exercise reasonable skill and care in the performance of his or her duties. This standard is also required of the designer by the Supply of Goods and Services Act 1982 in respect of any contract for the supply of design services. This statutory duty can be displaced by the imposition of a stricter duty in a contract.

The stricter duty is normally what is known as 'fitness for purpose'. Such a duty may be expressly stated in a contract in those words, or words to the same effect, or the law will usually imply it where the contract is on the basis of work and materials, unless it is clear that the employer is not relying on the contractor: *Young & Marten* v. *McManus Childs* (1968).

Where an employer relies on a contractor to provide an entire building and there is no independent designer involved, a term of reasonable fitness for purpose will be implied irrespective of any negligence or fault or whether the unfitness results from the quality of work or materials or from defects in the design: *Viking Grain Storage Ltd* v. *T H White Installations Ltd* (1985). The suggestion that matters of design in such circumstances should be regarded as involving no more than reasonable care was rejected by the House of Lords in the television mast case, *Independent Broadcasting Authority* v. *EMI Electronics Ltd and BICC Construction Ltd* (1980):

> 'As they undertook responsibility for the design, they became contractually responsible to ITA for BICC's negligence, and so in my opinion are liable to ITA in damages for breach of contract. In the circumstances it was not necessary to consider whether EMI had by their contract undertaken to supply a mast reasonably fit for the purpose for which they knew it was intended and whether BICC had by their contract with EMI undertaken a similar obligation but had that been argued, I would myself have been surprised if it had been concluded that they had not done so.'

A fit for purpose design liability might sometimes be implied into a designer's contract. *Greaves & Co (Contractors) Ltd* v. *Bayham Meikle & Partners* (1975) is a case in point, which dealt with the liability of an engineer carrying out the design of a warehouse floor for a contractor who was engaged by the employer on a design and build basis. Lord Denning said:

'The law does not usually imply a warranty that he will achieve the desired result but only a term that he will use reasonable skill and care. The surgeon does not warrant that he will cure the patient. Nor does the solicitor warrant that he will win the case. But, when a dentist agrees to make a set of false teeth for a patient, there is an implied warranty that they will fit his gums, see *Samuels* v. *Davis* (1943).

What then is the position when an architect or an engineer is employed to design a house or a bridge? Is he under an implied warranty that, if the work is carried out to his design, it will be reasonably fit for its purpose or is he only under a duty to use reasonable skill and care? This question may require to be answered some day as a matter of law. But, in the present case I do not think we need answer it. For the evidence shows that both parties were of one mind on the matter. Their common intention was that the engineer should design a warehouse which would be fit for the purpose for which it was required. That common intention gives rise to a term implied in fact.'

In that case, fitness for purpose was not implied as a matter of law but as a matter of fact, i.e. both parties had intended it to be so. In a situation where a JCT traditional standard form is being used, the contractor will have no design liability unless expressly stated to be part of the contractor's designed portion (CDP): *John Mowlem & Co Ltd* v. *British Insulated Callenders Pension Trust Ltd* (1977):

'I should require the clearest possible contractual condition before I should feel driven to find a contractor liable for a fault in the design, design being a matter which a [designer] is alone qualified to carry out.'

Such a condition is, of course, available in a limited way for CDP work in the JCT Standard Building Contract (SBC), the Intermediate Building Contract with contractor's design (ICD) and the Minor Works Building Contract with contractor's design (MWD).

3.1.4 Contractor's responsibility to warn

The question often arises as to the contractor's duty to warn the employer or the architect if it finds defects in the design. Considering the JCT Standard Form, 1963 edition, which requires the contractor to notify the architect if it finds any discrepancy in or between the documents, it was held that although the contractor has a duty under that form to notify discrepancies, it has no duty to look for and find the discrepancies in the first instance: *London Borough of Merton* v. *Stanley Hugh Leach Ltd* (1985). In the Canadian case of *Brunswick Construction Ltd* v. *Nowlan* (1974), a contractor charged with carrying out the construction of a house to architect's designs was held to be liable for an error in the design where the original architect was not engaged to inspect the work. In this as in other cases the question of reliance appears to be important.

In *Equitable Debenture Assets Corporation Ltd* v. *William Moss* (1984), it was held that to give efficacy to the contract, a term was to be implied requiring the contractor to warn the architect of design defects and that there was a duty of care in negligence to the employer and to the architect in this regard. The point was emphasised in *Victoria University of Manchester* v. *Hugh Wilson & Lewis Womersley and Pochin (Contractors) Ltd* (1984):

> 'In this case, I think that a term was to be implied in each contract requiring the contractors to warn the Architects as the University's agents of the defects in design, which they believed to exist. Belief that there were defects required more than mere doubt as to the correctness of the design, but less than actual knowledge of errors.'

The point was given further consideration in *University of Glasgow* v. *W Whitfield and John Laing Construction Ltd* (1988), where it was held that the decisions in *EDAC* and *Manchester* were concerned with the duty of a contractor to warn the employer, not with any duty owed by the contractor to warn the architect. Both cases assumed a special relationship of reliance between the contractor and the employer. Some doubt may have been thrown on this decision, however, when Judge Newey QC, who was responsible for the decisions in *EDAC* and *Manchester*, returned to the theme in *Edward Lindenberg* v. *Joe Canning & Jerome Contracting Ltd* (1992):

> 'In *Brunswick Construction* the Supreme Court of Canada held that experienced contractors had acted in breach of contract in building a house in accordance with plans prepared by an engineer in which they should have detected defects. In *Equitable Debenture* I held that there was an implied term in a contract requiring contractors to inform their Employer's Architect of defects of design of which they knew and in *Victoria Manchester University* I held that the implied term extended to defects which they believed to exist.'

In Lindenberg, the court held that an ordinary competent builder would have been expected to question suspected design defects. This is a point of some importance in the light of any gaps perceived to exist between the design obligation owed by the original architect and the contractor under DB. In *Bowmer & Kirkland* v. *Wilson Bowden Properties Ltd* (1996), the court observed that it was 'a feature of good workmanship for a contractor to point out obvious errors, or if there is doubt or uncertainty in the plans, specification or other instructions, to ask for clarification so that the uncertainty is removed'. It seems that, where there is any danger of injury or death, the contractor's duty to warn may have to amount to a refusal to carry on until the defect has been corrected: *Plant Construction plc* v. *Clive Adams Associates and JMH Construction Services Ltd* (2000).

3.2 *Liability under the contract*

The most important difference between DB and SBC is the contractor's obligation to design as well as construct. Virtually all the other differences spring from this central obligation. The design obligation deserves careful scrutiny. It is to be found in article 1 and clauses 2.1 and 2.17.

3.2.1 Article 1

The contractor is to complete the design for the Works and carry out and complete the construction of the Works. The Works are defined as the Works briefly described in the first recital and specifically referred to in the contract documents and including any changes made to the Works. Importantly, it is to be noted that the contractor is not to design, but to complete the design.

3.2.2 Clause 2.1 Contractor's obligations

The contractor's obligation under clause 2.1 is split into a number of distinct parts. First it must carry out and complete the Works in a proper and workmanlike manner in accordance with the contract documents. This would be the contractor's normal obligation under traditional forms such as SBC or IC. Its second obligation follows. The contractor is to complete the design of the Works.

The contractor is to complete the design for the express purpose of carrying out and completing the Works in accordance with the contract documents. The contractor is apparently not given responsibility for the design as a whole, but merely to complete what is, presumably, left incomplete. That, indeed, is what was generally understood until the decision in *Co-operative Insurance Society Ltd* v. *Henry Boot Scotland Ltd* (2002). This was a case which concerned the JCT 80 form of contract with amendments 1, 2 and 4–14 all as amended by the Contractor's Designed Portion Supplement. Although this was not a design and build contract, the clauses added by the CDP had many of the characteristics of the design and build contract, significantly including an obligation on the contractor under clause 2.1.2 to 'complete the design for the Contractor's Designed Portion . . .'. Several of the other clauses were virtually identical to the equivalent clauses in the design and build contract. Judge Seymour QC said:

> 'In my judgment the obligation of Boot under clause 2.1.2 of the Conditions was to complete the design of the contiguous bored piled walls, that is to say, to develop the conceptual design of [the engineers] into a completed design capable of being constructed. That process of completing the design must, it seems to me, involve examining the design at the point at which responsibility is taken over, assessing the assumptions upon which it is based and forming an opinion whether those assumptions are appropriate. Ultimately, in my view, someone who undertakes, on terms such as those of the Contract (that is to say including clause 2.7) an obligation to complete a design begun by someone else agrees that the result, however much of the design work was done before the process of completion commenced, will have been prepared with reasonable skill and care. The concept of *"completion"* of a design of necessity, in my judgment, involves the need to understand the principles underlying the work done thus far and to form a view as to its sufficiency. Thus I reject the submission of [counsel for Boot] that all Boot had to do in any circumstances was to prepare working drawings in respect of the bored pile walls. If and insofar as the walls remained incomplete at the date of the Contract, Boot assumed a contractual obligation to complete it, quite apart from any question of producing working drawings.'

Clause 2.7 is virtually identical to clause 2.17.1 of DB. The judge accepted the submission of counsel for the engineer:

> 'Boot's obligation to design the piles and the contiguous pile wall would necessarily have included a duty to check the adequacy of any preliminary design by others whether they were adopting or not any part of that preliminary design.
>
> Boot had the right to develop any design for the piled walls provided they complied with the required load-bearing capacities and complied with the minimum requirements for deflection.'

This is a most important judgment. It makes clear that the contractor's obligation to complete the design under the former CDP supplement extends to an obligation to check the design which has already been prepared to make sure that it works. Because the clauses in the CDP supplement closely reflected the clauses in WCD 98, it is thought that the judgment applies to WCD 98 also. This is good news for employers, but rather bad news for contractors. In drafting the new form, DB, the JCT have taken this judgment into account, as they expressly state in the *Guide*. An attempt has been made to restrict the contractor's responsibility to merely completing the design. This has been done by the inclusion of clause 2.11, which provides that the contractor is not responsible for the Employer's Requirements, nor for verifying the adequacy of any design contained in the Employer's Requirements. This appears to deal with the contractor's problem as raised by the *Co-operative Insurance* case. Whether it does so completely or whether there remain some loose ends is a matter for some future court. It is noteworthy that clause 2.11 is made subject to clause 2.15, which deals with divergences from statutory requirements. Therefore, the contractor cannot hide behind clause 2.11 if it finds a divergence from statutory requirements in the Employer's Requirements.

It appears to follow from clause 2.11 that if the employer, for whatever reason, has caused the whole of the design to be prepared by an independent architect and included within the Employer's Requirements, there will be no design to complete and the contractor's obligation in this regard will be non-existent. At the other end of the same spectrum, if there is no design included in the Employer's Requirements, the contractor will be responsible for the whole of the design.

The scheme of the contractor's obligations is clear. It is to construct the whole of the Works, but only to design whatever is left undesigned by the Employer's Requirements. Because most contracts let on this particular form include some design input on behalf of the employer, this clause immediately introduces the very thing which the philosophy of design and build is intended to avoid: uncertainty regarding design responsibility.

The second part of clause 2.1.1 is almost a repetition of what was in WCD 98 clause 2.1. It is ambiguous with at least two possible meanings:

- The contractor must complete the design of the Works only to the extent that it is not described or stated in either the Employer's Requirements or the Contractor's Proposals and the design is to include the selection of any specifications for any kinds and standards of the materials and goods and workmanship to be used in the construction of the Works; or
- The contractor must complete the design of the Works and must include the selection of any kinds and standards of the materials and goods and workmanship to be used in the construction of the Works if they are not described or stated in the Employer's Requirements or the Contractor's Proposals.

It is not absolutely clear whether it is the design, or the selection of any specifications, which may be in the Employer's Requirements or the Contractor's Proposals. Common sense suggests that it is the first meaning which was intended by the JCT in drafting this clause. However, at best it seems to state the obvious and, at worst, to throw doubt on the precise extent of the contractor's obligations under this clause.

3.2.3 Clause 2.17.1 Contractor's design warranty

This clause actually sets out the contractor's design liability. From the contractor's point of view, it is a most important clause and a good reason why a contractor should always opt for this contract rather than a simple exchange of letters, under which the contractor would be under a 'fitness for purpose' liability unless the lower standard was specified.

Under Clause 2.17.1 the contractor is to have the same liability to the employer as an architect or appropriate professional designer holding him or herself out as competent to do the design. The clause goes further and, lest there be any doubt, makes clear that it is referring to an architect or designer acting independently under a separate contract with the employer, having supplied the design for a building to be carried out by a contractor who is not supplying the design. The liability to which this clause refers is the liability of an averagely competent professional who is liable only to the extent that he or she fails to exercise reasonable skill and care (see section 3.1.2 above) and it is a mystery why it did not say so instead of continuing to use the present clumsy form of words.

This should be contrasted with the liability of a contractor that the building which it has designed and built is reasonably fit for any purpose which has been made known to the contractor (see section 3.1.3 above). The contractor's liability under this form of contract is considerably less than the liability it would shoulder at common law. To that extent this contract represents a very valuable restriction of the contractor's design liability. In view of the negotiated status of this contract, this restriction is not one which falls under the provisions of the Unfair Contract Terms Act 1977.

The liability is expressed to be the same as for a professional whether under statute or otherwise. In other words, the liability to the employer is the same as for any professional person under the Supply of Goods and Services Act 1982, section 13, reasonable care and skill, and is to be the same professional standard of care in tort also. It is unlikely that a court would find a higher standard of care applicable in tort where the parties had the opportunity to express such a standard in their contract and they had not taken it (see section 3.1.1 above). The contractor's liability is in respect of any defect or insufficiency in the design.

The clause very precisely sets out the boundaries of the design to which this professional design liability extends and they may be listed as follows:

- The design comprised in the Contractor's Proposals.
- The design which is to be completed in accordance with the Employer's Requirements.
- The design which is in accordance with the conditions, including any further design which the contractor is to carry out as a result of a change instruction.

At first sight, this provision covers all the design work which the contractor may do other than what has been carried out by or on behalf of the employer and incorporated in the Employer's Requirements. It seems, therefore, that if there is any design which the contractor is to carry out under the contract, but which can be brought outside the boundaries listed above, the contractor will have the ordinary liability of fitness for purpose in respect of such design.

3.2.4 Clause 2.17.2

This clause makes clear that if the contract involves any work in connection with dwellings, the contractor has liability under the Defective Premises Act 1972. The key provision of the Act is to be found in section 1(1) which provides: 'Any person taking on work for or in connection with the provision of a dwelling . . . owes a duty to see that the work which he takes on is done in a workmanlike or, as the case may be, professional manner, with proper materials and so . . . that the dwelling will be fit for human habitation when completed.' The obligation put on professionals should be noted. The contractor's design obligation under clause 2.17.1 is said to be that of an independent professional designer, and it therefore amounts to the duty to use reasonable skill and care except when dwellings are involved, when the Act bites and the obligation becomes, to all intents and purposes, to design so as to be fit for purpose.

Prior to 31 March 1979, dwellings sold with the National House Building Council (NHBC) guarantees were excluded from the Act, since the NHBC scheme was then approved. The NHBC scheme ceased to be approved after 31 March 1979 and consequently the Act applies to all persons taking on work for or in connection with the provision of a dwelling. As will be noted, the standard required by section 1(1) equates broadly to that required by the corresponding duties at common law.

The 1972 Act did not have retrospective effect (*Alexander* v. *Mercouris* (1979)). In *Thompson* v. *Clive Alexander & Partners* (1992), it was emphasised that the reference to fitness for habitation in section 1(1) merely sets the standard required in the performance of the duty created by section 1(1) and is not aimed at trivial defects.

Section 1 was considered by the Court of Appeal in *Andrews* v. *Schooling* (1991). In that case the plaintiff, the long lessee of a flat, claimed damages for breach of duty under section 1 on the grounds that the converted flat was not fit for human habitation. The flat suffered from damp caused by the evaporation of moisture from the cellar.

The Court of Appeal held that section 1 applied to both acts of commission ('misfeasance') and omission ('non-feasance') and that it was irrelevant that the problem did not become apparent until after completion. It is probable that the 1972 Act will become increasingly important in light of the restrictive decision of the House of Lords in *Murphy* v. *Brentwood District Council* (1990), especially as the NHBC scheme is no longer approved and there is no other approved scheme.

3.2.5 Clause 2.17.3 Limit of the contractor's liability

This clause gives opportunity for the employer to impose a limit on the contractor's liability in certain circumstances. An overriding proviso is that limitation on liability can apply only so far as it does not concern the contractor taking on work in connection with dwellings. That is to say, if the contract is for the design and construction of 12 flats, there can be no limitation of liability. If, however, the contract is for the design and construction of 40 houses and a school, liability could be limited in respect of the school.

In so far as dwellings are not involved, the employer may stipulate the limit of the contractor's liability for loss of use, loss of profit or other consequential loss arising in respect of the liability referred to in clause 2.17.1, i.e. design liability. This is a protection for the contractor and in appropriate cases the employer may state a limit in the hope that the reduced risk will be reflected in the tender prices. Alternatively, such a limit may be the result of negotiation following a high initial tender or as part of the usual two stage tendering procedure. The wording of the clause makes clear that the limit is to be stated in the contract particulars and that if no amount is therein stated, there will be no limit on liability under these heads. It seems, therefore, that if the employer states in the Requirements that there will be a limit on the contractor's liability under this clause, of £25,000, the contractor's liability will still be unlimited if the employer neglects to insert the amount in the contract particulars. This is because of the operation of clause 1.3 giving precedence to the agreement and the conditions.

The effect of this provision is probably rather less in practice than might appear to be the case at first sight. It really protects the contractor only 'from claims for special damages which would be recoverable only on proof of special circumstances and for damages contributed to by some supervening cause': *Saint Line* v. *Richardsons, Westgarth & Co Ltd* (1940). If a limit is stipulated in the contract particulars, it has no effect on the employer's power to deduct liquidated damages under clause 2.29 for failure to complete the Works by the completion date, and the deduction of such damages has no effect on the limit.

3.2.6 Clause 2.37 As-built drawings

Clause 2.37 provides that the contractor must provide certain documents before practical completion of the Works or a section. The documents are described as the contractor's design documents and related information describing the Works as built and relating to the maintenance and operation of the Works and any installations included in the Works. These documents must be as specified in the Employer's Requirements or as the employer may reasonably require. Therefore, it appears that if the documents are noted as required in the Employer's Requirements, the contractor's obligation will be to provide those documents. If nothing is noted for such documents in the Employer's Requirements, the contractor must provide what the employer reasonably requires. However, it does not appear to be open to the employer to insert requirements in the Employer's Requirements and, additionally, to request other documents.

By stipulating that the documents must be provided before practical completion, the contract is effectively saying that the contractor must provide as-built drawings before the Works have been built or at least completed. The reason for requesting the drawings before practical completion is presumably to enable the employer to deal with any maintenance or operational items arising immediately after occupation. It would have been logically more acceptable to have stipulated that the drawings should be provided within 2 weeks of practical completion, so that there would be no excuse for providing drawings which were not entirely accurate. A workable compromise might be to require the as-built documents in draft before practical completion and the final versions to be provided no later than 2 weeks later.

3.2.7 Clause 2.38 Copyright

Copyright in all the contractor's design documents is stipulated, by clause 2.38.1, to remain with the contractor. This is stated to be subject to any rights in designs provided to the contractor by the employer. Obviously, there will also be the rights of any architects or other designers engaged by the contractor as sub-contractors for design work, but these rights are not mentioned. The provisions of DB cannot override whatever may have been agreed between the contractor and its sub-contractors who were not a party to DB. Most sub-contract architects and designers will be unwilling to relinquish the copyright on their respective designs. This may give a problem to the contractor, but it is sensible to interpret clause 2.38.1 as setting out the position between the parties to DB so that, as between employer and contractor, the copyright belongs to the contractor except for any designs provided by the employer.

Clause 2.38.2 proceeds to grant to the employer an irrevocable, royalty-free and non-exclusive licence to use and to copy the contractor's design documents and to reproduce the contents for a whole string of purposes relating to the Works, including construction, repair, promotion, etc. To achieve this, the contractor will have to have obtained a similar licence from the sub-contract designers with the ability to grant sub-licences. Where the contractor uses Design and Build Sub-Contract Conditions DBSub/C, that is achieved by clause 2.25.

The grant of a licence to the employer is made subject to all money due and payable under the contract having been paid. This is a very sensible proviso from the contractor's viewpoint, but it could pose problems for the employer. Whether any amount is due and payable will depend on the contractor's application for payment, the employer's notices (if any) and, ultimately, on the provisions of clause 4.8. If it is found, perhaps after dispute resolution procedures have been invoked, that the employer has failed to pay money due and payable, the employer will no longer have a licence to reproduce the contractor's and its sub-contractors' designs and copyright will have been infringed from the date when the money should have been paid. Presumably the situation can be remedied, for the long term, by immediate payment, but one might speculate that the contractor could pursue a claim for damages for infringement for the intervening period. In common with most copyright clauses, the employer is not given a licence to reproduce the designs for extension to the Works.

The provision that the contractor is not to be liable for any misuse of the designs is probably otiose. The contractor could be liable only if the likelihood of such misuse was reasonably foreseeable: *Introvigne* v. *Commonwealth of Australia* (1980).

3.3 Design liability optional arrangements and consequences

The architect's role was briefly discussed in chapter 1.3. The architect and other members of the design team may be employed either by the employer or by the contractor. Although they may be either in-house or independent, for simplicity's sake it is assumed that the consultants are all independent. The position of the in-house designer, where different, is discussed further in section 3.6 below. Besides or instead of the architect, there may be several other consultants employed by the contractor. It is helpful to examine liabilities in three parts:

- Where the consultant carries out the whole of the design, or completes the design, for the contractor.
- Where the consultant produces the first part of the design, including the Employer's Requirements, for the employer.
- Where the consultant acts for the employer to obtain tenders, advises on the best tender to accept, puts together the contract documents and acts as employer's agent until completion of the project.

3.3.1 Consultant acting entirely for the contractor

In the first instance a consultant acting for the contractor will owe such duties to the contractor as are expressly noted in whatever contractual agreement has been executed between them. The architect's terms are briefly considered in section 3.3.4 below. It is probable that the consultants also owe the contractor a duty of care in tort. Where DB is used, each consultant will be a sub-contractor to the contractor under the provisions of clause 3.3.2. The contractor's liability to the employer will be the obligation to carry out the design using reasonable skill and care and the consultant should have a similar liability under an agreement with the contractor. Where the contractor employs a consultant who employs one or more sub-consultants, each sub-consultant is likely to have a duty in tort to the contractor in accordance with *Hedley Byrne* principles: *Cliffe Holdings* v. *Parkman Buck Ltd and Wildrington* (1996).

If a different kind of design and build contract is used (such as ACA 3), the contractor will probably have the higher liability of fitness for purpose and the contractor must ensure that the consultant has a similar level of liability. The problem here is that the consultant's professional indemnity insurers are virtually certain to reject any suggestion that the consultant takes on this higher level of liability and, without insurance, the higher liability has little practical value if a large claim is involved. Contractors must be wary of this point. In practice, the consultant's ordinary standard of liability is normally all that is required provided the consultant is made aware of all relevant criteria before commencing the design. As in the case of *Greaves & Co (Contractors) Ltd* v. *Bayham Meikle & Partners* (1975), it is difficult for an engineer to deny liability for a floor which is not fit for the purpose made known before the design was carried out. If, on receipt of the information, the engineer designed the floor using the same standard of reasonable skill and care to be expected from an ordinary competent engineer used to doing similar work, it should be fit for purpose.

Preparing design and constructional drawings for a contractor may be rather confusing at first for a consultant who has no experience of the design and build method of procurement. As noted above, the precise duties will depend principally on the terms of engagement. In general, however, some principles can be stated. The consultant is only empowered to do what is expressly agreed in the conditions of engagement. This may seem a very basic point, but to a consultant used to working in a traditional situation for the employer, it may come as a shock. For example, the contractor's design obligation is to complete the design and in doing so satisfy the Employer's Requirements. How far this obligation is transferred to the consultant is a matter for the contractor. If the Requirements ask for a heating system capable of producing a certain level of temperature under certain conditions, the contractor nevertheless may require the consultant to design a system giving more or less than this temperature or the contractor may set a totally different set of criteria to be satisfied. The consultant's duty is to satisfy the criteria communicated by the contractor, which may have its own reasons for amending parts of the Employer's Requirements. The consultant's duty is owed to the contractor, not to the employer.

If the consultant is the architect charged with overall design work on behalf of the contractor, similar considerations apply. If the architect becomes aware that the employer has requested certain changes under the provisions of clause 5, the architect has no authority to incorporate the results of such changes in the design unless the contractor expressly passes on instructions to that effect. That is not to say that if a consultant becomes aware of such matters, the contractor should not be informed. That should be done to protect the architect's position. The contractor may simply have made a mistake. But once notified, the consultant's duty to the contractor in respect of that particular matter is complete unless and until the contractor gives further instructions. Difficult questions may arise if the contractor asks the consultant to include in production information matters which are contrary to the Building Regulations or to good practice. Although the consultant must take the contractor's instructions, the consultant may not do anything contrary to law and there may be other circumstances where the consultant may deem it better to terminate under the conditions of appointment. Professionals cannot just blindly obey instructions. They have a broad and obvious duty not to act against their own professional judgment. Any professional in doubt about that should simply consider whether he or she would feel comfortable explaining such actions in any subsequent arbitration.

What architects and other consultants find difficult to accept is that they have no duty to the employer in these matters. Their professional skills are at the service of the contractor. Another situation which frequently arises is when the architect has detailed certain things on the drawings and possibly in a specification which has been produced for the contractor to enable construction to proceed on site. The contractor may indicate to the architect that it fully intends to construct the detail by another method and the architect may consider the contractor's method vastly inferior to the one that has been detailed. The architect certainly has a duty to point out to the contractor the shortcomings of the contractor's method, but if the contractor insists on doing things its own way or even ignores the architect's letter, the architect's duty has been discharged and no more can be done. The architect is not entitled to go directly to the employer. Of course, any consultant who is regularly ignored may decide that continuing to work under these circumstances is untenable. Whether the consultant is entitled to terminate the employment for that reason depends on the terms of the agreement.

These problems are fairly simple to resolve provided the consultant remembers that design and construction information is being prepared for the benefit of the contractor, not for the benefit of the employer. Such an idea frequently goes against the grain for consultants unused to working for contractors, but in fact there is nothing unprofessional or unlawful in it. The employer may have some of the design produced as part of the Employer's Requirements. In such a case, and despite the decision in *Co-operative Insurance Society Ltd* v. *Henry Boot Scotland Ltd* (2002), it would be wise for the architect engaged by the contractor to be satisfied that the initial design was workable. The architect is not simply entitled to accept the initial design as correct. Doubtless, the contractor will expect a consultant to identify such shortcomings before the Contractor's Proposals are formulated if the consultant is engaged at that point. Unless there is something clearly wrong with the initial design or the Requirements, it is very unlikely that the consultant has any duty simply to point out areas where the design might be improved.

Of course the consultant, as a member of the human race, will owe some general duties in tort to the employer and to others. A consultant whose design results in injury or death to a person, or if it causes damage to property other than the item designed, may be liable in tort of negligence. The consultant owes this duty to anyone who might foreseeably be affected by the design: *Murphy* v. *Brentwood District Council* (1990).

3.3.2 Consultant action for employer for first part of design

The liability of a consultant engaged by the employer to produce the first part of the design, including the Employer's Requirements and probably application for planning permission, is exactly the same as the liability of any consultant performing a partial service for a client. The consultant's duties should be set out in the conditions of engagement. In general, the consultant must use reasonable skill and care in taking instructions from the employer and in expressing those instructions in the form of Employer's Requirements. They should be so framed that a contractor tendering on the basis of satisfying those Requirements will produce the kind of Proposals envisaged by the employer. This is no mean task and the consultant must take great care in recording the employer's brief in the first instance. A consultant who fails to draft the Employer's Requirements with sufficient care may be liable to the employer for functional inadequacy in the resultant building if sufficient link between drafting and inadequacy can be established.

3.3.3 Consultant acting entirely for employer

If the consultant is engaged by the employer in the third instance, to deal with tendering, contract documentation and to act as agent under the contract, the duties will depend on the terms of engagement. There will normally be a duty to take reasonable skill and care in performing those services. While acting as employer's agent, the consultant will be governed by the normal law of agency (see section 5.1). The contract gives the agent power to act for the employer for the receiving or issuing of applications, consents, instructions, notices, requests or statements and otherwise to act for the employer under any of the clauses. That is, unless the employer specifies to the contrary by written notice to the contractor. A consultant filling this role is in a somewhat different position to the architect under SBC. The agent has no duty to the employer to act fairly between the parties, neither does the agent owe such a duty directly to the contractor. Thus in no sense is the consultant fulfilling an independent function. The wording of the contract makes the position clear: *J F Finnegan Ltd* v. *Ford Sellar Morris Developments Ltd (No 1)* (1991). There is no provision for any form of certification or other discretionary activity by the agent. Indeed, except for two clauses (clauses 2.7.3 and 3.1) the employer's agent is not specifically mentioned in the conditions.

A consultant must be scrupulous in acting within the authority given by the employer under the contract. It is quite common for an employer to wish to reserve certain functions and in some instances the agent may have few powers under the contract. A difficult situation may arise if the consultant acting as agent during operations on site is not the consultant who assisted the employer in the formulation of the Employer's Requirements. The second consultant may disagree with parts of the Requirements, more particularly if an initial sketch design forms part of the Requirements. Clearly, the second consultant must advise the employer if the objections to the Requirements are serious. Ideally, an employer wishing to engage a consultant purely for the purpose of obtaining tenders and acting as agent during the progress of the work should give the prospective consultant an opportunity to examine the Employer's Requirements before accepting the commission.

3.3.4 Architect's terms of engagement

Until recently, the terms of engagement for architects published by the RIBA were SFA/99 and CE/99, with various supplements including amendments (DB1/99 and DB2/99) said to be suitable for use by architects carrying out work for an employer or a contractor in a design and build situation.

The RIBA are about to publish new documents which are intended to provide the components of a flexible system which can be assembled to create contracts tailored to the needs of the particular project. They will be available in hard copy or electronic format and the idea is that, other than the conditions, the documents can be edited and re-branded as required. Although there are variants intended for domestic projects, it is the CA-S-07-A version which will probably be used with contract DB and replace SFA/99 and CE/99.

In order to complete an agreement, the architect must put together a number of different documents, some of which may be optional and some of which may be specially generated for the project in hand. So far as design and build projects are concerned, the set of documents forming the agreement is likely to consist of:

- *The Memorandum of Agreement:* This is the basic agreement for signature, which incorporates whichever documents are listed under item 3. A letter of appointment may be used instead.
- *The RIBA Standard Conditions of Appointment:* These are the terms of engagement which are considered below. Part A is said to be suitable for all clients while part B is only for use with public authorities and business or commercial clients.
- *Project Data:* This is intended to include all the variable parts of the agreement, some of which used to be part of the memorandum of agreement in SFA/99, but it also includes details of cost and consultant appointments.
- *Schedules of Fees and Expenses:* Although this is said to be optional, it is clear that the architect will require some agreed record of the fees and expenses to be charged and the intervals between payment.
- *Design Services Schedule:* This is supposed to be capable of use where the building contract is DB. It may be desirable to amend the specified services to suit particular requirements.

- *Role Specifications:* This is a set of specifications for a number of typical roles undertaken by the architect or consultant, for example, lead consultant, employer's agent, lead designer, CDM co-ordinator, etc. Further roles may be added if desired and, of course the existing specifications may be amended.

It is difficult to know, at this stage, what the average architect or other consultant in practice will make of these new documents. The use of a number of amendable documents to be assembled to suit a project and carefully completed with the addition of specifically tailored portions for different circumstances, sounds excellent in theory. However, the first impression is one of enormous complexity even if the second impression is more favourable. One suspects that most architects actually prefer a fairly simple and straightforward set of booklets such as SFA/99, CE/99 and SW/99 which, with minimal effort, could be used separately or with a supplement (e.g. DB1/99) for all occasions. Anecdotal evidence suggests that many architects have difficulty in completing the current forms where the choices to be made are not great. Many architects still operate on the basis of a simple letter of engagement and this set of documents is unlikely to get them to change.

Space does not allow for a detailed analysis of the RIBA Standard Conditions of Appointment, but previous users of SFA/99 should be aware and wary of some significant changes. The first very noticeable difference is that the conditions are mostly written in the present tense. This is similar to the approach under the NEC3 Engineering and Construction Contract noted in section 1.4.3. There is the same potential for ambiguity. Although it is usually possible to make a reasonable guess as to the meaning, a reasonable guess is what a legally binding agreement is intended to avoid. For example, where the Client 'gives decisions and approvals as necessary . . .' it is not clear whether the client has the duty to (shall) give them or simply the power (may), or whether the client has already given them or will do so in the future. To confuse matters, some clauses do include the word 'may'. It is difficult to understand the advantage of this mode of expression.

Under clause A2.5.3, the client is given power to require the removal of any of the architect's staff by exercising *the client's* reasonable opinion and, under clause A2.6.1 at the completion of services, the client can demand all documents provided to the architect without the safeguard of being made subject to all fees paid in full.

Definitions of 'Construction Cost' and 'Relevant Cost' are apt to be confusing. When read with clause A.5.3, which deals with the calculation of percentage fees, the result is something short of the clarity which is needed in this crucially important area. It is not abvious why 5% has been inserted as the amount to be added to the current Bank of England rate for late payment in clause A5.13. Why is it not 8% to match the Late Payment of Commercial Debts (Interest) Act 1998 rate? A client who pays on time does not have to worry about interest – whatever the rate. Hence, an architect should be wary of a client who complains about a high rate of interest for late payment. The provisions for payment on termination of clause A5.15 restrict the architect's ordinary common law right to damages on repudiation of the appointment.

Previous RIBA terms of engagement have contained the right for either architect or client to terminate at will, merely giving reasonable notice and stating the reason. This has always been a valuable clause to architects, particularly when an awkward client may make it impractical for the architect to continue although it may not be easy to identify a specific breach. Clause A8.4 retains the right for the client, but not for the architect. On the other hand, clause A8.5 removes the architect's common law right to accept the client's conduct as repudiation and bring the architect's own obligations to an end.

Although part B is apparently included to comply with statutory requirements, it omits any reference to notice which must be served by the client under section 110 of the Housing Grants, Construction and Regeneration Act 1996. Therefore, the Scheme for Construction Contracts (England and Wales) Regulations 1998 will apply and introduce an unwanted further complexity to an already complex document.

The overall content of the conditions is disappointing and the reader could be forgiven for being surprised to learn that these are RIBA terms.

3.4 Consultant switch and novation

Something has already been said about this topic in section 1.3 and there is further information in section 6.1. The system whereby a consultant acts first for the employer and later for the contractor is commonly, and often mistakenly, called 'novation'. Whereas 'consultant switch' involves the consultant entering into terms of engagement with the employer and then into a completely different contract with the contractor when the first comes to an end, novation occurs where a contract between employer and consultant is replaced by a contract between contractor and consultant on identical terms. It is perhaps easier, although inaccurate, to visualise novation as taking away one party to a contract and replacing with another. Novation requires an agreement between all three parties, but the major problem is that the contractor will not want the same terms as the employer (for example, it will not require the same services). Therefore, if it is to be effective, the novation must make provision for a change in the terms.

The benefit of novation is supposed to be that the consultant is made liable for all the design, even for early design carried out directly for the employer, and that this liability is owed to the contractor. Unfortunately, the protagonists of novation forget that the contractor is only liable under DB for *completing* the design, irrespective of whether the consultant is liable to the contractor for the whole design. Therefore, the employer's attempt to channel all design responsibility through the contractor will fail unless DB itself is fundamentally amended. An interesting liability situation may arise.

In theory, the system promotes a smooth design process because it simply continues with the same design team involved. But, as can be seen from section 3.3 above, while in contract with the employer, the consultants owe a duty to the employer to take reasonable skill and care in preparing the design and in giving advice in the best interests of the employer. In consultant switch situations, after the design team enter into contracts with the contractor, their duty in respect of further design and any related advice is owed to the contractor in the context of the contractor's reasonable profit expectations. In the case of novation, the duty owed by the consultant to the employer is wholly transferred to the contractor. The consultant and the employer owe each other no further duties. The contractor assumes the employer's duties to the consultant. Thus, the consultant has the same design obligation, but owed to different parties at the two stages.

In addition, there are different obligations to advise during the stages. The danger is that the employer, and sometimes the consultant, will forget that in the second stage the consultant owes no advisory duty to the employer. Thus, if the contractor instructs the consultant to change part of the design, the consultant must comply because the consultant is now acting for the contractor, even if the consultant knows or believes that the employer does not want that particular change. This is because the contractor has merely sub-let the design to the consultant and as between the employer and the contractor, the contractor has the design responsibility.

The consultant has generally only contracted to carry out the contractor's instructions regarding the design. These instructions may well be that the consultant must complete the design in accordance with the Employer's Requirements, but not necessarily so. A consultant's long-established client may find it hard to accept that the consultant is no longer looking after the client's best interests. Each project demands individual consideration and often expert advice if consultant switch or novation is contemplated. The consultant who is asked to become involved in design and build on the basis of consultant switch or novation should take some time to explain these points to the employer before accepting the commission. Indeed, since the first edition of this book was written, anecdotal evidence suggests that consultants acting first for the employer and then for the contractor encounter considerable difficulties and the best advice to consultants and to employers is to avoid these situations and act for one or the other party exclusively. However the arrangement is managed, the consultant is always placed in a position of possible, and often actual, conflict. It is difficult to envisage anyone other than construction professionals allowing themselves to get into this kind of situation.

It also goes without saying, or should do, that a consultant engaged by the employer to prepare the Employer's Requirements should not take an engagement to work for a contractor until after the tendering process is complete, nor should there even be a tacit understanding. To do otherwise opens the consultant to the charge of partiality when the contractors invited to tender pose the inevitable questions. In practice, many consultants carry on a dual role with little regard for acceptable professional conduct.

From the employer's point of view, consultant switch and novation can ensure a continuity of design, therefore less possibility of mistakes. It is not unusual, although inadvisable, for an architect to carry out initial design for the employer and to subsequently complete that design for the contractor while at the same time carrying out duties as employer's agent. It need hardly be said that such an arrangement will almost certainly give rise to a conflict of interest so far as the architect is concerned: *Blyth & Blyth Ltd* v. *Carillion Construction Ltd* (2001). It is not possible for the architect to keep such interests apart.

A further danger is that the employer might be encouraged to think of the architect as an architect under a traditional procurement situation rather than as agent with limited authority and duties. For example, when acting for the contractor in completing the design drawings, etc., the architect's duties as agent do not extend to informing the employer of instructions received from the contractor to amend certain parts of the design. Neither can the architect take instructions regarding the design directly from the employer. In such a situation, the employer must instruct the contractor who, in turn, should instruct the architect. If the contractor chooses not to so instruct the architect, the architect may not carry out the employer's instructions, no matter that he or she knows that they have been given. The contractor may have very many reasons why it decides not to give instructions to the architect. Clearly, architects should not agree to act for both parties in this way.

The case of *Blyth & Blyth Ltd* v. *Carillion Construction Ltd* (2001) highlighted some problems which can arise when the parties enter into a novation agreement. The claimants were consulting engineers who were claiming fees for professional services. The defendant was a contractor which counterclaimed against the engineers for losses suffered as a result of alleged breaches of contract. The interest of the case centres around the counterclaim. The contract between the employer and the contractor was the JCT 81 design and build contract, but the conclusions of the court are relevant to the DB contract. An important clause was inserted into the contract as follows:

> 'any mistake, inaccuracy, discrepancy or omission in . . . the design contained in the Employer's Requirements . . . shall be corrected by the Contractor but there shall be no addition to the Contract Sum in respect of such correction or in respect of any instruction of the employer relating to any such mistake, inaccuracy, discrepancy or omission.'

In addition, article 2 was changed so as to place responsibility on the contractor for any design of the Works which had already been carried out. There were a number of heads of counterclaim, but the court took as an example the claim for additional costs arising from alleged inaccuracies in the information provided as part of the Employer's Requirements and other information provided prior to tendering, which resulted in the contractor having to supply additional materials for which it could not claim additional payment from the employer because of the amendments to the contract noted above. The engineer's terms of appointment contained a clause permitting the employer to require the engineers to enter into a novation agreement in a form annexed to the appointment. In due course the novation agreement was executed by the three parties. The novation provided that the engineers' liability, whether before or after the novation, would be to the contractor, just as though the contractor had been named in place of the employer. There was a further provision by which the engineers agreed that their services would all be treated as having been performed for the contractor and *agreed to be liable to the contractor for any breach of the appointment which occurred before the date of the novation*.

The question was whether the contractor could claim against the engineers for loss caused to the contractor due to the engineers' alleged breach of their obligation to the employer to provide accurate information for the Employer's Requirements. The court decided that the contractor could not make that claim. Essentially, the court's reasoning was that the contractor could not claim against the engineers for pre-novation breach of duty owed to the employer unless the employer had suffered the losses. In respect of the pre-novation breaches, the novation agreement allowed the contractor to act as though it had been the employer at that time.

The court was uncertain whether a proper novation agreement was employed in that case. This was because the novation agreement and the appointment provided for the engineer to continue to owe duties to the employer and for the employer to be able to take over the appointment again under certain circumstances. It should be mentioned here that many so-called novation agreements drafted on behalf of employers still endeavour to place continuing duties on the consultants to the employer. Not only does this prevent the agreement becoming a true novation, it throws up serious concerns about the ability of the consultants to perform their duties adequately. Even without such complications, novation agreements pose many problems. In the *Blyth* case an attempt was made by the contractor to give effect to the novation agreement by substituting the word 'contractor' for 'employer' in the appointment whenever it occurred. This produced results which the court described as 'nonsensical'. As the court said:

> 'In my view, the difficulty is not simply textual but also reflects an underlying tension between on the one hand the designer's duties to the employer and on the other the conflict of interests between an employer and a contractor.'

3.5 Novation agreements

Because novation is the replacing of one unique contract with another on the same terms, but with one party substituted for another, the written agreement which the parties execute is usually especially drafted for each situation. Having said that, there is no reason why a standard novation agreement should not be used provided that it can be amended to suit individual circumstances. Such agreements are not things to be left to amateur draftspersons.

The structure of this kind of novation agreement is broadly as follows:

- The names of the three parties concerned.
- The recitals identifying the Works concerned, the terms of the appointment document, the terms of the building contract and any other matter believed to be relevant.
- The agreement including, among other things:
  - A clause releasing the consultant from any liabilities to the employer.
  - A clause releasing the employer from any liabilities to the consultant
  - A clause indicating the current situation regarding the consultant's fees
  - A clause setting out liabilities between consultant and contractor as though the contractor had been a party to the appointment instead of the employer from the beginning
  - A clause referring to agreed changes to the appointment, usually by reference to an attached schedule or appendix.
- The attestation where all parties execute the agreement as a deed. It is essential to complete the agreement as a deed, because if it is completed as a simple contract, there must be sufficient consideration passing between the parties. That is sometimes difficult to demonstrate. A deed does not require consideration to be binding on the parties.

Two standard novation agreements which are worth consideration are the following.

Novation Agreement CIC/NovAgr first edition 2004

This agreement is produced by the Construction Industry Council. There are detailed guidance notes available. It is expressly stated to be 'For use where the appointment of a consultant is to be novated from an employer to a design and build contractor'.

Although under clause 1, the employer discharges the consultant from *further* performance of the consultant's duties and the consultant releases the employer from further performance, there is no discharge from any liability which may arise from what has already been done by consultant or employer. Moreover, the consultant is not discharged from any obligation to provide warranties at the employer's request nor from any confidentiality obligation under the appointment provided that the latter obligation does not conflict with any duty now owed to the contractor following the novation. It is possible to envisage circumstances where the consultant will have some difficult decisions to make on the confidentiality issue. It seems therefore, on the basis of the *Blyth* case, that whether this agreement is a true novation agreement is questionable at best.

The agreement includes in clause 2 an essential provision which enables the parties to vary the services still to be performed by reference to an attached schedule on which the changes must be set out. Without this provision, the consultant may be left with an obligation to carry out services which are entirely inappropriate in relation to a contractor client. The employer has an obligation to pay any fees outstanding at the date of the agreement.

Clause 4 attempts to deal with the effects of the *Blyth* case by including a warranty to the contractor from the consultant that the pre-novation services, for which the contractor may be responsible under the main contract, have been carried out in accordance with the appointment. Clause 4(b) goes on to provide that the consultant will not be absolved from liability to the contractor simply because the employer has suffered no loss. This provision does not appear to protect the contractor for losses sustained in the *Blyth* circumstances.

Standard Form of Novation Agreement (20844941.05)

This novation agreement has been produced by the City of London Law Society. Detailed guidance notes are available.

This is a true novation agreement because it does not contain any provisions continuing old, or setting up new, consultant obligations to the employer. It provides that the employer and the consultant have no further duties one to the other, and that the relationship of consultant and contractor is as if the contractor had always been a party to the appointment instead of the employer.

The agreement does not include any provisions to allow the parties to vary the existing appointment terms, but the guidance notes make reference to the need to review the appointment document and to amend it as required so as to exclude services which cannot possibly be regarded as having been novated to the contractor. That would involve the parties agreeing amendments to the appointment and incorporating them in the existing appointment before the novation took place. That seems an unnecessarily complicated process. The CIC approach, novating the appointment but with a schedule showing the amendments, is to be preferred.

3.6 In-house or sub-let

From a liability standpoint, it makes little difference whether the contractor's design input comes from its own in-house design department or from independent consultants especially engaged for the project, or from a combination of the two. There are a few practical points to note, however. A contractor who engages independent consultants must take care that the terms of engagement with them are back to back with its liabilities under the main contract DB. It is not just sufficient to refer to DB in the terms of engagement. Even where no amendments have been made to the main contract clauses, there will always be the variable parts of the contract, the articles and the contract particulars, of which the consultant must have knowledge when entering into an agreement with the contractor. In many cases there will also be amendments to the contract itself and the consultant must have reference to all the terms of the main contract in the consultant's appointment. A particular point which the contractor should watch concerns professional indemnity insurance. It is essential that each consultant has insurance appropriate to the risks to be undertaken, bearing in mind the terms of the main contract. Ideally, each consultant should provide the contractor with documentary evidence to this effect from the insurers.

Where design is to be carried out in-house, the contractor has vicarious liability for the actions of its employees carried out in the course of employment: *Century Insurance Co Ltd* v. *Northern Ireland Road Transport Board* (1942). An architect or engineer in this situation is liable to the contractor for negligence in precisely the same way as any other employee. The fact that, in practice, professionals in employment do not carry personal professional indemnity insurance will probably ensure that it is not worthwhile for the contractor to sue them. In such cases, the contractor will have to carry its own professional indemnity insurance. The professional employees will want to know that there is a waiver of the insurer's subrogation rights.

3.7 Consultant warranties

Section 3.1.1 above set out the basic principles of liability. It is clear that a designer's liability in tort for negligence may be severely restricted. In order to overcome the problem, it is common for the employer to ask for collateral warranties (or duty of care agreements) from anyone with whom the employer is not in direct contract. In theory, such warranties should not be generally required in the case of a design and build contract, because all liability is gathered under the contractor's wing. However, part of the design may have been done for the employer and be incorporated in the Employer's Requirements and the contractor has no liability for it. In addition, it is always possible that the contractor may go into liquidation. In these instances, it is in the employer's interests to enter into collateral agreements with all the consultants and with any other sub-contractors.

DB now provides for sub-contractor warranties which can be applied to consultants (see section 6.6) or the employer may require entirely separate warranties. The terms of the warranties will normally provide, at the very least, that the warrantor will use reasonable skill and care in carrying out any design function, selection of materials and the satisfaction of any performance specification. Where consultant switch is operated, separate warranties are a great advantage to the employer, because any design fault is the responsibility of the consultant whether engaged directly by the employer or through the medium of the contractor. This is not the place to discuss the wording or content of warranty forms in detail and the parties should obtain expert advice. Matters commonly covered by warranties include the following:

- Design
- Selection of materials
- Satisfaction of performance specification
- Professional indemnity insurance
- Assignment
- Copyright
- Deleterious materials
- Notice if termination in prospect and provision for novation
- Dispute resolution.

It remains to be seen whether collateral warranties are really effective or necessary. In the case of *Beoco Ltd* v. *Alfa Laval Co Ltd* (1994), a sub-contractor's collateral warranty was held to impose liability for all loss in respect of a weld badly done by the defendant sub-contractors, although on the facts the defective weld was not the cause of the damage.

Chapter 4
The Contractor's Obligations

4.1 *Express and implied terms*

The contractor will have obligations in respect of both express and implied terms. This is something which is seldom appreciated by employers or contractors, who tend to consider that the contract documents represent the whole of the terms governing the agreement. An express term is one which the parties have agreed; in contrast, an implied term is one which is written into the contract as a matter of law. The doctrine of implied terms is of the greatest practical importance.

4.2 *Implied terms*

An implied term is one which was not expressed in writing or orally at the time the contract was entered into, but which will be implied into contracts in a number of instances:

- By statute, for example the Sale of Goods Acts 1979 and 1994, the Supply of Goods and Services Act 1982 and the Housing, Grants, Construction and Regeneration Act 1996 set out several terms which will be implied into contracts which fall within the scope of those Acts.
- Where the parties have agreed on language with a particular meaning: *The Karen Oltmann* (1976).
- By custom or trade usage: *Symonds* v. *Lloyd* (1859).
- By common law. In *Liverpool City Council* v. *Irwin* (1977), the House of Lords noted two distinct circumstances in which the courts might imply terms:
  - — If it were necessary to give business efficacy to a contract
  - — If the terms were simply spelling out what both parties knew was part of the bargain.
- Other grounds for implying a term were stated in *Mackay* v. *Dick* (1881), namely 'where in a written contract it appears that both parties have agreed that something shall be done, which cannot be done unless both concur in doing it, the construction of the contract is that each agrees to do all that is necessary to be done on his part for the carrying out of that thing, though there may be no express words to that effect'.

Even in a very detailed contract such as DB, it is possible that it will be necessary to imply terms, as happened in the case of *Merton* v. *Leach* (1985) in a contract in the then current JCT 63 form.

The courts will not do so to improve a contract, only if it is essential to enable the contract to work. The following are terms which are commonly implied into building contracts:

- The contractor will carry out its work in a good and workmanlike manner – that is, it will carry out the work using the same degree of skill as would an averagely competent contractor who is experienced in that kind of work: *Hancock and Others* v. *B W Brazier (Anerley) Ltd* (1966).
- The contractor will supply good and proper materials – that is, materials which are satisfactory for their purpose: *Young & Marten* v. *McManus Childs* (1968).
- The contractor undertakes that a building will be reasonably fit for its purpose if that purpose has been made known to the contractor and if there is no other designer involved: *Viking Grain Storage Ltd* v. *T H White Installations Ltd and Another* (1985)
- The contractor will complete the Works by the date for completion or, if no date is stated in the contract, within a reasonable time of being given sufficient possession of the site: *Fernbrook Trading Co Ltd* v. *Taggart* (1979).

These terms may be modified or superseded by the express terms of the contract, which are normally given preference. Even in the traditional form of contract on SBC terms where an independent architect has been retained to prepare detailed drawings and specification and to administer the contract during operations on site, it has been found necessary to imply terms to make the contract commercially effective.

A useful discussion of the implication of terms in a specially-drafted 'design and construct' turnkey contract, for the provision of a floating production and storage facility in a North Sea oilfield, is to be found in the case of *Davy Offshore Ltd* v. *Emerald Field Contracting Ltd* (1992).

In contracts based on DB, however, the situation is likely to be somewhat different. There may well be significant gaps in specification between the essentially performance specification part of the Employer's Requirements and the perhaps less than exhaustive treatment of the operational specification part of the Contractor's Proposals. In such instances, it is clear that the law will imply appropriate terms to deal with workmanship and materials. The question of fitness for purpose is dealt with in detail in Chapter 3. Suffice to say here, that a term of this sort will be superseded by the provisions of clause 2.17. The contractor's powers and duties are listed in Figs 4.1 and 4.2, respectively. The remainder of this chapter deals with the most important of the contractor's duties.

4.3 Express terms

4.3.1 Contractor's obligations

The contractor's obligations are set out in what are probably the most important provisions of the contract:

- Article 1
- Clause 2.1.

Clause	Power	Comments
2.2.1	Substitute materials, goods and workmanship for those described in the Employer's Requirements or the Contractor's Proposals or specification.	Only if the original goods etc. are not procurable, and the employer consents in writing.
2.5.1	Consent to the employer using or occupying the site of the Works before the issue of the practical completion certificate.	If insurers confirm that insurance will not be prejudiced, consent must not be unreasonably withheld.
2.7.1	Inspect the Employer's Requirements and the Contractor's Proposals.	These documents are kept by the employer.
2.9.2	Consent to the carrying out of such work by others.	If the employer so requests where the Employer's Requirements do not so provide. Consent must not be unreasonably withheld.
2.30	Consent to the employer taking partial possession before practical completion.	
3.3.1	Sub-let all or part of the Works.	With the written consent of the employer.
3.3.2	Sub-let all or any part of the design.	With the written consent of the employer.
3.8	Request the employer to specify in writing which contract provision empowers the issue of an instruction.	
3.9.1	Consent to a change in the Employer's Requirements which is or makes necessary any alteration or modification in the design of the Works.	Consent must not be unreasonably delayed or withheld.
4.1.9	Write to the employer stating that it has incurred or is likely to incur direct loss and/or expense and quantify the same, not reimbursable under any other contract provision.	If it becomes or should reasonably have become apparent that regular progress has been and is likely to be affected by one or more of the specified matters.
4.11	Suspend performance of obligations.	If the employer fails to pay in full by the final date for payment and the contractor has given a 7-day notice of intention to suspend. The contractor cannot suspend if the employer has properly served a notice under clause 4.10.4.
4.16.2	Request the employer to place retention in a separate bank account.	

Fig. 4.1 Contractor's powers under DB (including powers and duties in the schedules).

5.2	Agree with the employer the valuation of changes and provisional sum work.	
7.1	Assign the contract.	With the written consent of the employer.
8.9.1	Give notice to the employer specifying the default.	If employer fails to pay by final date for payment; *or* fails to comply with assignment provisions; *or* fails to comply with the CDM Regulations.
8.9.2	Give notice to the employer specifying the suspension events.	If the Works are suspended for a period stated in the contract particulars due to any impediment, prevention or default of the employer.
8.9.3	Give notice terminating its employment.	If the default or suspension is not ended within 14 days.
8.9.4	Give notice terminating its employment.	If the default or suspension event is repeated.
8.10.1	Give notice terminating its employment.	If the employer becomes insolvent.
8.11.1	Give 7-day notice of termination.	If the Works are suspended for a period stated in contract particulars due to: • *force majeure*; *or* • certain employer's instructions; *or* • loss by specified perils; *or* • civil commotion, terrorist activity; *or* • UK government statutory power; *or* • development control requirements.
9.3	By written notice jointly with the employer to the arbitrator to state that they wish the arbitration to be conducted in accordance with any amendments to the JCT 2005 CIMAR.	
9.4.3	Give a further notice of arbitration to the employer and to the arbitrator referring another dispute which falls under article 8.	After an arbitrator has been appointed. Rule 3.3 applies.
Schedule 2, para. 4.2.2	Raise reasonable objection to provision of estimates within 10 days of receipt of instruction.	May do so on behalf of any sub-contractor.

Fig. 4.1 *Continued*

Schedule 3, para. B.2	Unless the employer is a local authority – Require the employer to produce for inspection the insurance policy and the premium receipts. Insure in joint names all work executed etc. against all risks.	 Where the Works are to be insured in joint names by the employer. If the employer fails to take out insurance
Schedule 3, para. C.3	Unless the employer is a local authority – Request employer to produce receipt showing that an effective policy is in place under paras C.1 and C.2. Insure in joint names and for that purpose enter the premises to make an inventory and survey.	 Applies to existing structures where employer is to insure in joint names. If the employer fails to take out insurance.
Schedule 3, para. C.4.4	Terminate its employment under the contract if it is just and equitable to do so. Invoke dispute resolution procedures.	This must be done within 28 days of the occurrence of the loss or damage, and is effected by written notice to that effect served on the employer. If the employer serves notice terminating the contractor's employment; *and* the contractor alleges that it is not just and equitable to do so; *and* the contractor acts within 7 days of receiving the notice.

Fig. 4.1 *Continued*

Clause	**Duty**	**Comments**
2.1.1	Carry out and complete the Works in accordance with the contract documents, the construction phase plan and statutory requirements, and complete the design for the Works including the selection of any specifications for any kinds and standards of materials, goods and workmanship so far as not described or stated in the Employer's Requirements or Contractor's Proposals.	Unless varied under clause 9.
2.1.3	Pass statutory approvals to the employer when they are received.	
2.1.4	Comply with any instruction and be bound by employer's decisions under the contract.	

Fig. 4.2 Contractor's duties under DB (including powers and duties in the schedules).

2.2.1	Provide materials and goods of standards described in Employer's Requirements or Contractor's Proposals. Not to substitute materials or goods.	So far as procurable. If not so described. Unless employer gives written consent.
2.2.2	Provide workmanship of standards described in the Employer's Requirements or Contractor's Proposals or appropriate to the Works.	To extent not so described.
2.2.3	Provide samples of goods and workmanship as specifically referred to in the Employer's Requirements or the Contractor's Proposals.	Before carrying out work or ordering goods.
2.2.4	Provide the employer with reasonable proof that the goods comply with clause 2.2.	If the employer so requests.
2.2.5	Take all reasonable steps to encourage contractor's persons to be recognised under the Construction Skills Certificate Scheme.	
2.3	Begin the construction of the Works when given possession of the site. Regularly and diligently proceed with the Works and complete them on or before the completion date.	This is subject to the provision for extension of time in clauses 2.23–2.26.
2.5.1	Notify insurers under insurance option A, B or C and obtain confirmation that use or occupation will not prejudice insurance.	Before giving consent to use or occupation.
2.5.2	Notify employer of amount of insurance premium. Give receipt to employer.	Where insurance option A applies and insurers require an additional premium. If so required and use or occupation is still required.
2.6.1	Permit the execution of work not forming part of the contract to be carried out by the employer or by persons employed or otherwise engaged by the employer.	If the Employer's Requirements provide the contractor with such information as is necessary to enable it to carry out and complete the Works in accordance with the contract.
2.7.3	Keep available to the employer's agent at all reasonable times one copy of the Employer's Requirements, Contract Sum Analysis, Contractor's Proposals and documents referred to in clause 2.8.	

Fig. 4.2 *Continued*

2.8	Provide the employer free of charge with two copies of the contractor's design documents.	In accordance with the contractor's design submission procedure, the contractor must not commence work until the procedure has been observed.
2.10.2	Give the employer written notice immediately on finding any divergence between the Employer's Requirements and the definition of the site boundary.	
2.13	Give the employer written notice immediately, specifying any discrepancy discovered in the documents.	
2.14.1	Inform the employer in writing of the contractor's proposed amendment where there is a discrepancy within the Contractor's Proposals.	
2.14.2	Inform the employer in writing of proposals to deal with the discrepancy.	If Contractor's Proposals do not deal with a discrepancy within the Employer's Requirements.
2.15.1	Notify the employer immediately in writing on finding any divergence between the statutory requirements and the Employer's Requirements or the Contractor's Proposals. Inform the employer in writing of the contractor's proposed amendment for removing the divergence. Complete at its own cost the design and construction of the work in accordance with the amendment.	With the employer's consent, which must not be unreasonably delayed or withheld. This is subject to clause 2.15.2.
2.16.1	Supply such limited materials and execute such limited work as are reasonably necessary to secure immediate compliance with statutory requirements.	In an emergency and if this is necessary before receiving the employer's consent.
2.16.2	Forthwith inform the employer of the emergency and steps it is taking under clause 2.16.1.	
2.17.1	Design the Works using the same standard of skill and care as would an architect or other appropriate professional advisor holding him or herself out as competent to take on work for such a design.	
2.18	Pay all statutory fees and charges and indemnify the employer against liability in respect of them.	No adjustment is made to the contract sum unless such fees are stated as a provisional sum in the Employer's Requirements.

Fig. 4.2 *Continued*

2.19	Indemnify the employer against liability in respect of copyright, royalties and patent rights.	If the employer instructs the use of patented articles etc., the contractor is not liable in respect of infringement, and all royalties, damages, etc. are added to the contract sum (clause 2.20).
2.22	Not to remove listed items except for use on the Works.	If the value has been included in an interim payment.
2.24	Notify the employer in writing forthwith of the material circumstances (including the cause or causes of delay), identifying any event which in the contractor's opinion is a relevant event.	If and when it becomes reasonably apparent that the progress of the Works or a section is or is likely to be delayed. The duty is in respect of any cause of delay and is not confined to the relevant events specified in clause 2.26.
	Give particulars of the expected effects of any relevant event and estimate the extent, if any, of the expected delay to completion beyond the currently fixed completion date. Keep the particulars and estimate up to date by further written notices as may be reasonably necessary.	If practicable, this must be done in the above notice, but otherwise in writing as soon as possible thereafter.
2.25.6	Constantly use its best endeavours to prevent delay in progress to the Works or any section and to prevent the completion being delayed or further delayed beyond the completion date; *and* do all that may be reasonably required to the satisfaction of the employer to proceed with the Works or section.	The second part of this duty does not require the contractor to spend money.
2.29	Pay or allow the employer liquidated damages at the rate specified in the contract particulars.	If the contractor fails to complete the Works or any section by the completion date; *and* if the employer has notified it to that effect; *and* if the employer has given written notice not later than the date when the final account and final statement become conclusive that payment of liquidated damages may be required; *and* if the employer has given written notice requiring payment.
2.30	Issue a written statement to the employer identifying the part of the Works or section taken into possession and giving the date of possession.	If employer with contractor's consent takes possession of part of the Works.

Fig. 4.2 *Continued*

2.35	Make good at its own cost all defects, shrinkages and other faults specified in the schedule of defects.	The defect etc. must be due to the contractor's failure to comply with its contractual obligations. The employer's schedule of defects must be delivered to the contractor not later than 14 days after the expiry of the rectification period. The contractor must remedy the defects etc. within a reasonable time of receipt of the schedule.
	Comply with any instruction issued by the employer requiring the remedying of defects, shrinkages and other faults.	No such instruction can be issued after the delivery of the schedule of defects or after 14 days from the expiry of the rectification period.
2.37	Supply the employer with 'as-built' drawings free of charge.	This must be done before practical completion of the Works or section.
3.1	Allow the employer's agent and any person authorised by the employer access to the Works, workshops, etc. at all reasonable times and ensure a similar right of access in any sub-contract.	Subject to reasonable restrictions and protection of proprietary rights.
3.2	Keep a competent person in charge on site.	At all times.
3.5	Forthwith comply with all instructions issued by the employer.	The instructions must be in writing *and* expressly empowered by the contract. The contractor need not comply where the instruction requires a change under clause 5.1.2 to the extent that it makes reasonable objection in writing to the employer.
3.7.2	Confirm to the employer in writing any oral instruction issued.	
3.9.4	Notify the employer in writing whether it has any objection under Regulation 20 of the CDM Regulations.	If the contractor is the CDM co-ordinator, within a reasonable time after receipt of a change instruction.
3.12	Open up for inspection any work covered up and arrange for the testing of any materials or goods or executed work.	If the employer so instructs. The cost will be added to the contract sum unless the tests etc. are adverse.
3.13.1	Remove from site work, materials and goods not in accordance with the contract.	If the employer so instructs.

Fig. 4.2 *Continued*

3.15	Use its best endeavours not to disturb any fossils, antiques, etc. found on site and cease work if its continuance would endanger the object found or impede its excavation or removal. Take all necessary steps to preserve the object in the exact position and condition in which it was found. Inform the employer of the discovery and the precise location of the object.	
3.16	Permit the examination or removal of the object by a third party.	If the employer so instructs.
3.18	Duly comply with the CDM Regulations.	
3.18.2	Comply with all the duties of the CDM co-ordinator. Prepare and deliver the health and safety file under the CDM Regulations	While the contractor remains the CDM co-ordinator. Free of charge to the employer.
3.18.3	Ensure that the construction phase plan is received by the employer before construction work is commenced. Notify any amendment to the plan to the employer.	While the contractor is and remains the principal contractor.
3.18.5	Provide, and ensure any sub-contractor provides, information for the preparation of the health and safety file.	Where the contractor ceases to be the CDM co-ordinator.
3.19	Comply with all the reasonable requirements of the principal contractor.	Where the employer appoints a successor to the contractor as principal contractor and where the requirements are necessary for compliance with the CDM Regulations. Free of charge to the employer.
4.9	Apply for interim payment, accompanied by such details as may be stated in the Employer's Requirements.	Under alternative A the applications are to be made on completion of each stage. Under alternative B applications are to be made at the period stated in the contract particulars up to the date named in the practical completion statement or one month thereafter and thereafter as and when further amounts are due and on the latest of the expiry of the rectification period or the issue of the notice of completion of making good. The employer is not required to make any such interim payment within one calendar month of having made a previous interim payment.

Fig. 4.2 *Continued*

4.12.1	Submit to the employer the final account and final statement for the employer's agreement. Supply such supporting documents as the employer may reasonably require.	Within 3 months of practical completion.
4.19	Submit to the employer such information as is reasonably necessary to ascertain the amount of direct loss and/or expense.	Upon the employer's request.
6.1	Indemnify the employer against any expense, liability, loss claim or proceedings whatsoever in respect of personal injury to or death of any person.	If the claim arises out of or in the course of or is caused by the carrying out of the Works *unless* and to the extent that the claim is due to any act or neglect of the employer or of any employer's persons.
6.2	Indemnify the employer against any expense, liability, loss, claim or proceedings whatsoever in respect of injury or damage to any property.	In so far as the injury or damage arises out of or in the course of or by reason of the carrying out of the Works; *and* is due to the negligence, omission or default of the contractor or any contractor's persons. 'Property' does not include the Works or site materials and loss or damage by specified perils to property with clause schedule 3, para. C.1 insurance.
6.4.1	Maintain necessary insurances for injury to persons or property.	The obligation to maintain insurance is without prejudice to the contractor's liability to indemnify the employer.
6.4.2	Produce documentary evidence of insurance cover.	When reasonably required to do so by the employer who may (but not unreasonably or vexatiously) require production of the policy or policies and premium receipts.
6.5	Maintain in joint names of the employer and the contractor insurances for such amount of indemnity as is stated in the contract particulars for damage to property other than the Works caused by collapse, subsidence, etc. Deposit with the employer the policy(ies) and premium receipts.	Where it is stated in the Employer's Requirements that this insurance is required. Employer must approve insurers.
6.9	Ensure that the joint names policies referred to in schedule 3, para. A.1 or para. A.3 *either* • provide for recognition of each sub-contractor as insured; *or* • includes insurer's waiver of rights of subrogation.	If insurance option A applies. In respect of specified perils.

Fig. 4.2 *Continued*

6.10.4	With due diligence restore any work damaged, replace or repair any unfixed materials or goods that have been lost or damaged, remove and dispose of debris and proceed with the carrying out and completion of the Works.	If the employer does not terminate under clause 6.10.2.2 and work or site materials suffer loss or damage due to terrorism.
6.11.1	Take out PI insurance in accordance with contract particulars.	Forthwith after entering into the contract.
6.11.2	Maintain PI insurance for period in contract particulars.	Provided it remains available at commercially reasonable rates.
6.11.3	Produce documentary evidence that PI insurance has been effected and/or maintained.	As and when reasonably required by the employer.
6.12	Immediately give notice to the employer and discuss best means of protecting their respective positions.	If the PI insurance ceases to be available at commercially reasonable rates.
6.14	Comply with the Joint Fire Code and ensure its persons also comply.	
6.15.1	Ensure remedial measures are carried out.	If a breach of the code occurs and the insurers specify the remedial measures required. Copy notice to employer.
7E	Comply with the contract documents as to obtaining warranties in the form SCWa/P&T, SCWa/F, SCWa/E or from consultants.	Within 21 days of receipt of employer's notice where part 2 of the contract particulars provides for the giving of a warranty by the sub-contractor.
8.5.2	Immediately inform the employer in writing if it makes a proposal, gives notice of a meeting or becomes the subject of any proceedings or appointment relating to clause 8.1 matters.	
8.7.2	Remove from the Works any temporary buildings, plant, tools, equipment, goods and materials belonging to or hired to the contractor. Provide the employer with two copies of all contractor's design documents then prepared even if previously provided. Assign to the employer without payment the benefit of any sub-contracts etc. within 14 days.	As and when so required by the employer in writing. In the event of termination by the employer under clauses 8.4–8.6. So far as assignable and lawful.

Fig. 4.2 *Continued*

8.12	With all reasonable dispatch remove from site all temporary buildings, plant, tools, equipment, goods and materials belonging to the contractor and contractor's persons. Provide the employer with two copies of all 'as-built' documents then prepared. Prepare an account. Provide the employer with all documents necessary for the preparation of the account.	In the event of termination under clauses 8.9–8.11 etc. At the employer's option if terminated under clauses 8.11 or 6.10.2.2 or schedule 3, para. C.4.4. Not later than 2 months after the date of termination.
9.4	Serve on the employer a notice of arbitration.	If the contractor wants a dispute referred to arbitration.
Schedule 1, para. 1	Prepare and submit two copies of each of contractor's design documents.	In such format as stated in the Employer's Requirements or the Contractor's Proposals in due time.
Schedule 1, para. 5.1	Carry out the Works in strict accordance with the submitted document.	If marked 'A' by employer.
Schedule 1, para. 5.3	Take due account of employer's comments and either re-submit in amended form or notify employer under para. 7.	If marked 'C' by employer.
Schedule 1, para. 7	Notify employer in writing that compliance would give rise to a change. Give reason in accompanying statement. Amend and re-submit the document.	Within 7 days if contractor disagrees with employer's comments and considers that the document is in accordance with the contract. If the employer confirms the comment within 7 days.
Schedule 2, para. 1.2	Appoint a manager. Not to remove or replace the manager.	Prior to start of Works on site. The employer must have consented in writing. Without the employer's written consent which must not be unreasonably withheld or delayed.
Schedule 2, para. 1.3	Ensure that the manager attends meetings in connection with the Works.	As and when reasonably requested by the employer.
Schedule 2, para. 1.4	Ensure the manager keeps complete and accurate records and makes them available for the employer at all reasonable times.	In accordance with any provisions in the Employer's Requirements.
Schedule 2, para. 2.1.1	Enter into contract with named sub-contractor and notify employer of date.	

Fig. 4.2 *Continued*

Schedule 2, para. 2.1.2	Immediately inform the employer of the reasons.	If unable to enter into a sub-contract with a named sub-contractor.
Schedule 2, para. 2.1.5	First obtain consent of the employer.	If the contractor wishes to terminate the named sub-contractor's employment.
Schedule 2, para. 2.1.6	Complete any balance of sub-contractor's work.	If the named sub-contractor's employment is left unfinished.
Schedule 2, para. 2.1.7	Account to the employer for amounts recovered or which ought to have been recovered using reasonable diligence.	Reasonably due to the employer as a result of the termination.
Schedule 2, para. 2.1.8	Include a provision that the named sub-contractor will not contend that the contractor has suffered no loss.	In the named sub-contractor conditions.
Schedule 2, para. 4.2	Within 14 days of an instruction, submit para. 4.3.1–4.3.5 estimates to the employer.	If compliance entails valuation, extension of time, or loss and/or expense. The contractor need not submit the estimates if the employer within 14 days of the instruction gives written statement that they are not necessary; *or* if within 10 days the contractor makes objection.
Schedule 2, para. 4.4	Take all reasonable steps to agree the estimates with the employer.	After submission to the employer.
Schedule 2, para. 5.2	Submit an estimate of the amount of loss and/or expense incurred in the period immediately preceding.	With the next application for payment if the contractor is so entitled under clause 4.19.
Schedule 2, para. 5.3	Submit an estimate of loss and/or expense with each application for payment.	For so long as the contractor continues to incur loss and/or expense.
Schedule 3, para. A.1	Insure the Works in joint names against all risks for their full reinstatement value.	Applicable to new buildings – the obligation continues until the date of the practical completion statement or date of termination. The insurance is to be placed with insurers approved by the employer. Para. A.3.1 enables this cover to be by means of the contractor's all risks policy.
Schedule 3, para. A.2	Deposit the policy(ies) and premium receipts with the employer.	

Fig. 4.2 *Continued*

Schedule 3, para. A.4.1	Give written notice to the employer.	Forthwith upon discovering loss or damage caused by risks covered by the joint names policy.
Schedule 3, para. A.4.3	With due diligence restore any work damaged, replace or repair any unfixed materials or goods that have been lost or damaged, remove and dispose of debris and proceed with the carrying out and completion of the Works.	After insurance claim under option A and any inspection required by the insurers.
Schedule 3, para. A.4.4	Authorise insurers to pay insurance money to employer.	Acting also on behalf of sub-contractors recognised pursuant to clause 6.9.
Schedule 3, para. B.3.1	Notify forthwith the employer of the extent, nature and location of any loss or damage.	Upon discovering the loss or damage caused by risks covered by joint names policy in para. B.1.
Schedule 3, para. B.3.3	With due diligence restore work damaged, replace or repair any unfixed materials or goods that have been lost or damaged. Remove and dispose of debris and proceed with the carrying out and completion of the Works.	The restoration of damaged work etc. is to be treated as a change in the Employer's Requirements and valued under clause 5.
Schedule 3, para. B.3.4	Authorise insurers to pay insurance money to employer.	Acting also on behalf of sub-contractors recognised pursuant to clause 6.9.
Schedule 3, para. C.4.1	Notify forthwith the employer of the extent, nature and location of any loss or damage affecting the Works etc.	The contractor must do this upon discovering the loss etc.
Schedule 3, para. C.4.3	Authorise insurers to pay insurance monies to employer.	Acting also on behalf of sub-contractors recognised pursuant to clause 6.9.
Schedule 3, para. C.4.5.1	With due diligence reinstate and make good loss or damage and proceed with the carrying out and completion of the Works.	If no notice of termination is served or if the dispute resolution procedures, having being invoked, have decided against the notice of termination.

Fig. 4.2 *Continued*

Article 1 very simply and straightforwardly says that the contractor must complete the design and carry out and complete the construction of the Works in accordance with the contract documents. Clause 2.1 has been considered in section 3.2.

In order to properly understand those obligations, the definition of 'Works' contained in clause 1.1 should be studied. Not only are the Works briefly described in the first recital, they are also 'more particularly' in the contract documents. The definition is extremely broad, referring to the Works 'shown, described or referred to' in those documents. The definition could scarcely be wider. 'Works' also includes any changes under the contract. The contractor's obligation is absolute. There is no qualification. Its obligation ends only at practical completion of the Works or upon the operation of the termination provisions.

The reference to changes makes clear that reference to the Works also includes all changes, in other words, the Works in their final condition. At first sight, the contractor is undertaking to carry out and complete the whole of the Works, including all changes however numerous, for the contract sum in article 2 and within the contract period stipulated in the contract particulars. That indeed was a worry when considering the equivalent provisions of WCD 98, possibly only saved by the obligation being made subject to the conditions.

Even very onerous contracts have been enforced. A contract requiring the contractor to carry out the Works and certain variations which might be ordered within a period originally agreed even if extension of time is not given, has been held by the courts as binding on a contractor even though it might involve an impossibility: *Jones and Another v St John's College Oxford* (1870). Whether a modern court would reach the same decision is debatable.

The new wording is much better. The contractor's obligation to carry out and complete the Works in article 1 is no longer directly linked to the contract sum in article 2. Therefore, although the contractor's obligation remains an obligation to carry out the Works (which by definition includes any changes), the amount payable to the contractor under article 2 is the contract sum or such other sum payable under the terms of the contract. The revised wording in both articles ensures that the contractor is entitled to be paid the value of any change.

The principal difference between this and other contracts is of course the contractor's obligation to complete the design for the Works as well as carrying out and completing the construction of the Works. Note that the obligation in respect of design is that the contractor must complete the design. For example, if the design was presented to the contractor in the Employer's Requirements as fully completed, theoretically the contractor, thanks to clause 2.11, would have no design obligation remaining. If the employer simply presented the Requirements in the form of a performance specification and nothing more, the contractor's design obligation would be total. The design obligation is discussed in detail in Chapter 3. Clause 2.1.1 is very basic to the contract and very often, when the parties are bogged down in argument regarding respective liabilities, it is worth returning to this clause, which sets the rest of the contract in perspective. The second part of the clause is not as clear as it could be. After the obligation to complete the design of the Works, the remainder of the clause could mean either that it must not include the selection of any specifications for any kinds or standards of materials and workmanship if they are in the Employer's Requirements or the Contractor's Proposals, or it might mean that the design must be completed only to the extent that it is not in the Employer's Requirements or the Contractor's Proposals and that the specifications of standards of materials or workmanship are included in the design. On balance, the better view is that the second interpretation is correct.

4.3.2 Regularly and diligently

The contractor's duty to proceed regularly and diligently with the Works after being given possession of the site is to be found in clause 2.3. It is important, particularly in the context of the termination provisions. Failure to proceed regularly and diligently is grounds for termination (see Chapter 12). It is discussed in detail in section 7.2.

4.3.3 Workmanship and materials

The contractor's obligations in respect of workmanship and materials are included in clause 2.2. They are similar to the obligations under SBC, but there are some significant differences. Clauses 2.2.1 and 2.2.2 deal with materials and goods, and workmanship, respectively. The contractor's duty is to provide materials and goods as described in the Employer's Requirements. If they are not described in the Requirements, the materials and goods are to be as described in the Contractor's Proposals or in the documents referred to in clause 2.8, but only in so far as they are procurable.

This is a substantial protection for the contractor, because if it were not for this proviso, the contractor would have an absolute obligation to provide the goods and materials. That is to say, it would not be able to offer any excuse for failure to so provide. Of course, if this saving phrase were omitted, there would be situations arising in which the contractor would be unable to comply, because the item would be unobtainable anywhere at any price. In some circumstances, it may result in the contract being frustrated. In practice, the contractor's approach in those circumstances probably would be to try to suggest other materials for the employer's approval which are equal in every way to those specified, at no additional cost to the employer. The effect of the saving provision is that if the contractor cannot obtain goods and materials as described, its obligations in this respect are at an end. Of course, this does not provide an escape for a contractor who is finding it more difficult than expected to get materials at a reasonable cost. It is sometimes said that the contractor should be expected to order all materials and goods at the beginning of a contract, then this situation would not arise. If the employer had the right to expect the contractor to order all materials at the very beginning of the contract, the contractor ought to have a complementary right to deliver to site and to receive payment. Payment for materials off site would not be discretionary.

Clause 2.2.1 refers to two distinct situations. The first is where the kinds and standards of materials are described in the Employer's Requirements, and the second is where they are described in the Contractor's Proposals. In the former situation, it is the decision of the employer to require the particular item. In the second, it is the decision of the contractor to propose an item in response to a performance specification. Clause 2.2.1 is not specific about the point at which the materials are not procurable. For example, it does not say 'so far as procurable at the base date'. It seems that 'procurable' is referring to the point in time when the item is required for construction or any preliminary work. If that is correct, and there appears to be no other sensible approach, the clause is wider in effect than generally thought. The contractor's obligation to provide an item of a kind or standard in the Employer's Requirements or the Contractor's Proposals is at an end if it is not procurable or obtainable when it is required. There is no precise machinery in the contract to deal with the situation when materials are not procurable. Under the traditional provisions of SBC, the material which the contractor is unable to procure would be material specified by the architect and an architect's instruction would be necessary to change the materials. There appears to be nothing in DB which obliges the contractor to suggest an alternative item, whether at the same or greater cost, even if the item concerned was specified by the contractor itself in the Contractor's Proposals. Certainly, there is nothing to compel the contractor to absorb the extra cost of a replacement item. Realistically, the employer should embody a substitute material in a change instruction so that it can be valued and the contract sum can be adjusted. In such circumstances, the contractor may be entitled to an extension of time and loss and/or expense. It is noteworthy that clause 2.2.1 ends by prohibiting the substitution of materials and goods without the written consent of the employer. This provision refers to a situation where the contractor, for reasons of its own, wishes to change the material from the material described. The contract is silent regarding the mechanism to be employed if the contractor wishes to substitute. No doubt in many instances, the parties will deal with the situation in a fairly informal way. The contractor may well write to the employer suggesting the substitution of one material for another. An astute employer, realising that the contractor has made the suggestion because there is some advantage to be gained, may require the contractor to specify the saving in cost which the employer can expect if consenting. However, there is no mechanism to achieve the reduction in the contract sum which should result. The only way to achieve that would be for the employer to issue an instruction requiring a change, and so allow the substitution to be valued.

Unfortunately for that line of approach, the contract does not empower the employer to instruct a change in materials. Clause 5.1.1 refers to a 'change in the Employer's Requirements which makes necessary . . . '. Among the things made necessary are the 'alteration of the kind or standard of any of the materials or goods to be used in the Works'. Therefore, an employer cannot instruct a change in materials or goods in the Contractor's Proposals, but only a change in the Employer's Requirements which results in such a change. This precludes the direct instruction of a change in materials or goods unless they are expressly specified in the Employer's Requirements. For example, if the Employer's Requirements specify that XYZ marble should be used on the floors in the foyer, there is nothing to prevent the employer from issuing an instruction changing the marble to slate, hardwood or anything else. However, if the XYZ marble is specified in the Contractor's Proposals as a response to the Employer's Requirements which simply call for a good quality and prestigious floor finish, the employer has no power to directly instruct a change in the marble floor, but may only change the requirement, say to a hardwearing but relatively inexpensive covering, to which the contractor must respond by a change in material. Clause 4.1 makes clear that the contract sum can only be adjusted or altered in accordance with the express provisions of the contract.

A further example highlights the problem. Is the employer entitled to a reduction in the cost of the Works as a result of simply writing and agreeing to a drawing, submitted under clause 2.8, which shows the substitution of a cheap material for the more expensive one specified in the Employer's Requirements/Contractor's

Proposals? At first sight, again there appears to be no mechanism to deal with the situation, even in schedule 1, and it might be said that an employer who simply agreed without precondition has agreed to the reduction in quality without a corresponding reduction in price. But, if there is no contractual procedure to accomplish a price reduction following a substitution, still less can there be the procedure to reduce quality *without* a price reduction. A more correct view may be that, whether or not the employer stipulates a price reduction before agreeing to the substitution, there would have to be a reduction (i.e. a payment from the contractor to the employer) as the consideration for the employer's agreement. Since there is no mechanism for adjusting the final account and final statement to reflect the reduction, the adjustment would have to be made outside the contract as part of a collateral contract. This kind of side-agreement is relatively common in construction work, although often the parties are not aware of the precise legal relationship they are setting up and the variation is valued as though it had been *validly* instructed. That, of course, is the whole point. The obvious solution is to bring the situation within the change provisions so that it can be valued under the contract provisions. The matter is not without doubt and it is again hoped that JCT will direct their minds to solve the problem at the next amendment.

Under clause 2.2.4, if the employer so requires, the contractor must supply reasonable proof that the materials and goods comply with clause 2.2. It is not stated what form this proof should take. WCD 98 referred to 'vouchers'.

Clause 2.2.2 deals with workmanship. The standard of workmanship is to be as described in the Employer's Requirements. To the extent that the standard is not so described, it must conform with the Contractor's Proposals or the documents issued by the contractor under clause 2.8. It is notable that in this clause and clause 2.2.1, noted above, the Employer's Requirements are given precedence. For example, if a material or standard or workmanship is variously described in the Requirements and the Proposals, it is clear that the description in the Requirements is to be preferred. Only if there is no specific description in the Employer's Requirements, can the Contractor's Proposals be consulted.

Clause 2.1.1 requires the contractor to carry out all work in a proper and workmanlike manner and in accordance with the construction phase plan (see also section 5.3.2). Whether a contractor is complying with the first part of this duty in any particular instance will be a matter to be decided with reference to any term in the Employer's Requirements or the Contractor's Proposals, any relevant code of practice and the practice in the industry.

The provision of samples is required by clause 2.2.3. However, this is by no means an all-embracing clause as is sometimes thought by employers and their agents. The provisions are quite precise and amount to the following:

- In respect of workmanship, goods or materials,
- if samples of the standards or quality of such workmanship, goods or materials are *specifically* referred to in either the Employer's Requirements or the Contractor's Proposals, and
- if the contractor intends to provide such workmanship, goods or materials,
- the contractor must provide the employer with such samples before either carrying out the work or ordering the materials.

For example, the Employer's Requirements may state that 'the contractor must provide a sample of the floor finish one metre square, on appropriate base to show the standard of workmanship and the quality of materials'. If it is the Contractor's Proposals, probably the wording will be rather more precise because the contractor is essentially putting forward its specification for the work. On its true construction, it is considered that the clause must specify whether the sample is intended to show standard of workmanship or quality of materials or both. In some cases it will be obvious, for example where a sample of the type of brick or roof cladding is required. Not until the units are combined will any workmanship considerations apply. Generally worded admonitions in the Employer's Requirements, such as 'samples of all goods and materials intended for use on the Works must be provided to the employer before ordering', or 'samples must be provided to the employer as required from time to time', are not thought to be sufficiently specific to fall within the terms of this clause and they are void by operation of clause 1.3 giving priority to the printed form (see section 2.7). From a simple common sense point of view, the contractor will be unable to price for such vague provisions. The contract is silent on the position if the employer dislikes the sample presented. There is no approval procedure and the contractor is entitled to provide the required samples and then simply proceed with the Works. The clause does not state that the contractor must provide the samples at any particular time period before carrying out the work or ordering the materials, and the contractor would strictly comply with the clause if it provided them just the day before it was due to take action, although it may be prudent to allow a longer period. It is not thought that the courts would imply any term that a reasonable period should be allowed, in view of the fact that where the contract wishes some action to be reasonable, it so states. It also specifies particular periods in other cases, and the context of the contract as a whole does not require it: *R M Douglas Construction Ltd* v. *CED Building Services* (1985). This clause is clearly intended to enable the employer to check in certain circumstances, as the work progresses, that the contractor is providing what the contract documents specify. But it will inevitably happen that the employer dislikes something that was required, when there is the opportunity to actually see it. For this reason, the employer would be wise to amend the clause so that it stated a time period such as '5 working days', to give the opportunity to decide whether to change the requirements. Of course, any such change would be subject to clause 5 (Changes) and to clauses 2.23–2.26 (Extensions) and 4.19–4.22 (Loss and expense), if any delay or disruption was caused. The whole philosophy of this contract is to place responsibility on the contractor to satisfy the employer's carefully formulated requirements and, therefore, the employer should be wary of making expensive changes. This clause should be used for its primary purpose, to check that the contractor is complying with the terms of the contract.

Clause 2.21 provides that the materials and goods which are intended for the Works and which are placed on or adjacent to the Works but not fixed, may not be removed without the employer's consent in writing. The employer must not unreasonably withhold consent. For example, it would be unreasonable to withhold consent to the removal of materials which for one reason or another were no longer required as part of the Works or which had been the subject of repeated thefts. The second part of this clause attempts to provide that the materials and goods will become the property of the employer after their value has been included in any interim payment. This clause will be effective against the contractor, but it cannot transfer the ownership of the materials to the employer unless, at the time of the purported transference, they belong to the contractor. This clause is not effective against sub-contractors, sub-sub-contractors or suppliers, unless some similar provision is included in their contracts. This contract, like SBC, requires the contractor under clause 3.4 to insert such clauses into its sub-contracts. There is a similar clause in the standard form of sub-contract DBSub/C (clause 2.15). However, the provision would have to be inserted right down the contractual chain to overcome this problem: *Dawber Williamson Roofing Ltd* v. *Humberside County Council* (1979). Once the materials are incorporated into the Works, the ownership passes to the employer: *Reynolds* v. *Ashby* (1904).

Materials stored off site by the contractor are dealt with in clause 2.22 and 4.15. If the employer wishes to include the value of off-site materials in stage payments or in interim payments, they must be included in a list attached to the Employer's Requirements. The contractor must satisfy the list of seven criteria. These criteria are intended to protect the employer against the risk of paying for materials which the contractor does not legally own, and also to safeguard the employer if the contractor should become insolvent. In the latter situation, the employer can only recover the materials if it can be shown that there is no doubt that specific labelled materials belong to the employer (off-site materials are discussed at greater length in Chapter 10).

4.3.4 Statutory obligations

The obligations placed on the contractor by the contract are of the greatest importance in the context of its general design and build obligations. The obligations are no longer gathered together in one clause, but spread over a number of clauses in the contract. The key provision is clause 2.1.1, which states that the contractor must carry out the Works in compliance with statutory requirements. That means it must comply with all Acts of Parliament, instruments, rules or orders or any regulation or bylaw of any local authority or statutory undertaker with any authority in regard to the Works. The definition of statutory requirements is expressly stated to include development control requirements, which are defined elsewhere in clause 1.1 as any statutory provisions and any decision of a relevant authority thereunder which control the right to develop the site. In simple terms, the contractor must comply with the requirements for obtaining planning permission. Although no longer expressly stated, the obligation to comply with statutory requirements must include the obligation to submit any statutory notices. Under clause 2.1.3, the contractor must pass all approvals to the employer.

The contractor, therefore, must ensure that the Proposals comply with all local planning authority requirements and in this connection it is vital that the employer spells out in the Requirements just what is required. It is quite common for the employer to have obtained outline planning permission or even full planning permission while leaving the contractor to comply with conditions imposed by the planning authority. There is an important proviso in clause 2.1.2 that if the employer states in the Requirements that they comply with statutory requirements, the contractor has no duty under the contract to so comply nor to give any notices relating to the subject of the compliance. If the employer states that the Requirements comply partially, then to the extent that they comply, the contractor's obligations are reduced accordingly. Thus, the employer should make sure that copies of any applications and approvals sent and received are included as part of the Employer's Requirements. In practice, it is often unclear whether the employer has stated that the Requirements comply. In such circumstances, the contractor should seek written confirmation of the position at the time of formulating its Proposals before tendering.

Clause 2.15.2.2 provides that if the terms of any permission or approval of the planning authority after the base date stated in the contract particulars have the effect of amending the Contractor's Proposals, the amendment is to be treated as if it is the result of a change in the Employer's Requirements under clause 5. This situation could arise in several ways. The employer may obtain outline planning permission or even full planning permission with reserved matters and the contractor may produce its Proposals and tender on this basis. After the base date (which is normally a date of about the same date as the tender is submitted), the planning authority may give decisions on the reserved matters which are in conflict with what the contractor has included in its Proposals. In such circumstances, and provided the contractor has formulated its Proposals on the basis of information then available, it is entitled to have the necessary amendment to its Proposals treated as a change in the Employer's Requirements and reimbursement would follow. In another situation, the employer may not have obtained any planning permissions. When the contractor eventually receives permission, any amendments necessary to its Proposals will rank as if they were made necessary by a change to the Employer's Requirements. There is a proviso that the employer has not precluded such treatment in the Requirements. If it is so precluded, the contractor must stand the increased costs resulting from such amendments. An employer who seeks to place such total responsibility on the contractor will usually pay a heavy price in an increased tender figure. If the employer has specifically stated in the Requirements that they or any part comply with statutory requirements, any amendments to the Requirements in order to conform with statutory requirements are dealt with under clause 2.15.2.3 and must be the subject of a specific instruction of the employer requiring a change. If it was not for this contractual provision, the employer's statements in the Requirements could be held to be misrepresentations and the contractor would have its remedies under common law or under the provisions of the Misrepresentation Act 1967. Changes in statutory requirements after base date, requiring amendments to the Contractor's Proposals, are covered by clause 2.15.2.1 and they are to be treated as a change. The contractor is obliged to comply with statutory requirements even if the Employer's Requirements do not so comply. This must be the case, because the contractor's duty to comply with statutory requirements takes priority over its contractual obligation to satisfy the employer. In addition, clause 1.3 ensures that the requirements of clause 2.15.2 take precedence over the Employer's Requirements.

Clause 2.15.1 provides that if either the employer or the contractor finds a divergence between statutory requirements and the Employer's Requirements (including any change under clause 5) or the Contractor's Proposals, the finder must give immediate written notice to the other. Whoever gives the notice, the contractor must send the employer written proposals for removing the divergence. Provided the employer consents, and such consent may not be unreasonably withheld, the contractor must proceed to incorporate the amendment at its own cost. There is an important proviso that the amendment will not be at the contractor's cost if the problem is caused by one or more of the situations envisaged in clause 2.15.2 discussed above, which may be summarised as changes to statutory requirements after the base date or amendments necessary to the Employer's Requirements which are specifically stated to comply.

Clause 2.16 makes provision for any emergency compliance with statutory requirements. This will normally involve some danger to health and safety or imminent structural collapse, leaving no time for the contractor to seek specific instructions from the employer or the employer's agent. The contractor must supply the minimum necessary materials and carry out the minimum amount of work necessary to comply and overcome the emergency. The extent to which the contractor is entitled to be reimbursed will depend on the precise circumstances in accordance with the various principles set out above. As soon as reasonably practicable, the contractor must let the employer know of the problem and the steps being taken to overcome it.

Clause 2.18 deserves careful attention. As might be expected, it makes the contractor responsible for paying all fees and charges legally demandable under any statutory requirement in connection with the Works. The contractor must include for all such payments in its tender figure. Only if they are expressed as provisional sums in the Employer's Requirements will the contractor be entitled to receive the actual amount expended. Provisional sums are unlikely to be included for this purpose unless the figures are likely to be of such size and so unpredictable that the contractor will be obliged to include a very large sum in its tender figure to cover the risk. A point often overlooked is the indemnity which the contractor gives to the employer in respect of liability for the fees and charges. Its effect is that if the contractor fails to pay, it agrees to reimburse the employer, not only for the actual amount of the charges but also for any consequential losses without restriction. Its effect is much broader than the employer's normal remedy for the contractor's failure, which would be to sue for damages for breach of contract.

It is worthwhile considering the position of what are referred to in the contract as statutory undertakers. These are organisations such as the water supplier, gas suppliers and electricity suppliers which are authorised by statute to construct and to operate public utility undertakings. They derive their powers from statute either directly or through statutory instruments. They can be involved in the contract either in performance of their statutory obligations or as contractors or sub-contractors. In the performance of their statutory obligations, they are not liable in contract: *Willmore* v. *S E Electricity Board* (1957), but in certain circumstances they may have a tortious liability. It is possible that they are directly engaged by the employer under clause 2.6, or they may be sub-contractors under clause 3.3 or schedule 2, paragraph 2, or they may be statutory undertakers. It is important to establish in which capacity they are on site, because if they disrupt the regular progress of the work, the contractor will be entitled to an extension of time and loss and/or expense if they are considered to be employer's licensees under clause 2.6. The contractor will be entitled to an extension of time if they are simply acting in their capacity of statutory undertakers. If acting as named or domestic sub-contractors, the contractor must bear the risks of disruption itself unless it can recover its losses from the statutory undertakers.

The parties are required to comply with the CDM Regulations. By making compliance with the Regulations a contractual duty, breach of the Regulations becomes a breach of contract so providing both employer and contractor with remedies under the contract (clause 3.18). Clause 3.18.1 provides that the employer 'shall ensure' that, if the contractor is not the CDM co-ordinator, the CDM co-ordinator carries out all the duties of a CDM co-ordinator under the regulations and that, where the principal contractor unusually is not the contractor, the principal contractor will also carry out its duties in accordance with the regulations. If the contractor is the CDM co-ordinator, it must carry out all the duties of a CDM co-ordinator under the regulations (clause 3.18.2). There are also provisions that the contractor, if it is the principal contractor, must ensure that its construction phase plan is with the employer before work commences on site and that it notifies the employer of any subsequent amendment (clause 3.18.3).

In situations where the contractor is not the CDM co-ordinator or perhaps has ceased to have that responsibility, it must provide, and ensure that any sub-contractor provides, information reasonably required by the CDM co-ordinator. This must be done within the time reasonably required by the CDM co-ordinator and notified in writing to the contractor.

Every change instruction issued by the employer potentially carries a health and safety implication, which must be examined and the appropriate procedural steps taken under the regulations. The regulations present the CDM co-ordinator with very grave responsibilities. Some of those duties must be carried out before work is started on site. If necessary actions delay the issue of a change instruction or, once issued, delay its execution, the contract provides that the contractor is entitled to extension of time (clause 2.26.5) and any loss and/or expense it can substantiate (clause 4.20.5). Every construction professional should have a thorough grasp of the regulations and the contractual clauses which deal with them, so that the full consequences of any new instruction can be carefully considered before it is issued.

4.3.5 Person-in-charge

Clause 3.2 requires the contractor to keep on site a competent person-in-charge who must be on site 'at all times', i.e. during the whole of the time the Works are being executed. The contractor may designate anyone as person-in-charge and the idea is that there is always someone available on site to whom the employer can issue instructions, confident that such instructions are being issued to the contractor. It is, therefore, essential that the person-in-charge thoroughly understands the implications of clause 3.5 and, in particular, the need to get all instructions in writing before complying.

Schedule 2, paragraph 1 provides for the appointment of a site manager. Where these provisions apply, they replace clause 3.27. From the employer's point of view, it is very worthwhile applying paragraph 1 because it requires the contractor to obtain the employer's written consent before appointing the site manager and in the case of any change in the appointment. The employer may withhold or delay consent to removal or replacement of the site manager, but not unreasonably. It seems that there is no such requirement for reasonableness in respect of the initial appointment. The site manager is to be full-time on site and instructions given to the site manager are deemed to have been given to the contractor. Paragraph 1.3 requires the manager, together with the contractor's persons as necessary, to attend any meetings which the employer may convene in connection with the Works. The manager is also required, by paragraph 1.4, to keep complete and accurate records in accordance with any provisions which are included in the Employer's Requirements. Therefore, if there are no provisions concerning records in the Employer's Requirements, the manager has no duty to keep such records under this clause. In practice, of course, any competent manager will keep records for the contractor's benefit. Such records will contain details of weather, visitors to site, instructions given, deliveries, men employed and on which operations, progress, notable occurrences, and so on. If the employer has specified particular records in the Requirements, the manager must make them available for inspection by the employer or the employer's agent at all reasonable times, i.e. during normal working hours.

4.3.6 Instructions

Clauses 3.5–3.9 are vital to the proper execution of the contract. They are discussed in detail in section 5.3. The contractor's principal duties under these clauses are as follows:

- The contractor must comply with all instructions issued by the employer as soon as it reasonably can do so.
- This is subject to the contractor's right to query the empowering provision in the contract. The contractor may accept the employer's response and thereby receive all the benefits flowing from the instruction, whether or not it actually is empowered under the specified clause.
- The contractor need not comply if it makes reasonable objection to an instruction requiring the imposition of any obligations or restrictions or alterations to such obligations or restrictions in respect of access to the site, limitations of space or hours or the execution of the work in a specific order; or
- If the instruction results in an alteration to the design of the Works and with good reason, it does not consent; or
- If the contractor is the CDM co-ordinator and it has an objection under Regulation 14 of the CDM Regulations. If so, it must give written notification to the employer within a reasonable time of receiving the instruction. The employer must vary the instruction to remove the cause of the objection until the contractor is satisfied. The contractor may not continue its objection unreasonably; or
- If the instruction is oral. In this case the contractor must confirm it in writing to the employer within 7 days. If the employer does not dissent, the contractor's obligation to comply takes effect from the expiry of 7 days from receipt of the contractor's confirmation. The contractor need not confirm if the employer does so first. The contractor must comply from the date of the employer's confirmation. There is provision in clause 3.7.4 for the employer to confirm at any time before the conclusivity of the final account and final statement if neither party confirmed, but the contractor nevertheless complied with an oral instruction.

- If the contractor fails to comply with a written notice from the employer requiring compliance within 7 days, clause 3.6 empowers the employer to engage others to carry out the instruction and charge all costs in connection with the operation to the contractor.

4.4 *Other obligations*

4.4.1 Drawings

The contractor's obligations to provide drawings are found in clauses 2.8 and 2.37. Clause 2.8 stipulates that the contractor must provide to the employer without further charge two copies of the contractor's design documents (i.e. the drawings, specifications and details) which it prepares or uses for the Works. The drawings are for information only. Although there is no express reference to setting out dimensions and levels, it is clear that, despite the fact that the employer must define the site boundaries in accordance with clause 2.9, the contractor is responsible for setting out on site. That duty is to be implied as a vital part of the contractor's obligations under the design and build contract.

Clause 2.37 requires the contractor to supply the employer, before practical completion, with as-built drawings and other information showing operational and maintenance details. Its obligation is not open-ended, but merely to supply such contractor's design documents and other information as may be specified in the contract documents or as the employer may reasonably require

Clause 2.8 gives the contractor further obligations by reference to schedule 1 which makes provision for the submission of drawings to the employer. Alternatively, a procedure may be set out in the contract documents, probably in the Employer's Requirements. Schedule 1 is pretty straightforward, but it should be read carefully by the employer and the contractor to avoid any misunderstandings.

Schedule 1 provides a detailed procedure. In what amounts to a disclaimer, paragraph 8.3 makes clear that the contractor's obligation to ensure that the contractor's design documents are in accordance with the contract is not reduced by reason of compliance with the submission procedure in schedule 1, nor by any comment of the employer. The schedule refers to the 'Contractor's Design Documents' and the definition in clause 1.1 is wide enough to include any drawing, detail, specification or related documents. The contractor must submit two copies of each to the employer in the format set out in the Employer's Requirements or the Contractor's Proposals. If there are formats set out in both Requirements and Proposals, it is the Employer's Requirements which takes precedence.

The submission has to be made 'in sufficient time' to allow the contractor to amend the documents to include the employer's comments before using the documents in connection with the Works. The phrase 'in sufficient time' is something of a cop-out for the employer. One has only to consider the process described below and its possible variations to see that the contractor may have an impossible task if there are repeated submission of documents marked 'C'. It is suggested that the contractor would only have to demonstrate that it allowed a reasonable time in the light of all the circumstances known to it at the time of submission.

Paragraph 2 gives the employer 14 days from receiving a document from the contractor, or from the end of any period which may have been stated in the contract documents, to return one copy of the document to the contractor. The returned documents must be marked either 'A', 'B' or 'C'. The employer may only mark 'B' or 'C' if the documents are not in accordance with the contract. This is a most important point which is often overlooked by the employer or the employer's agent. The procedure is not the opportunity for the employer to have second thoughts about anything or to insist that the contractor constructs any part of the Works in a special way. For example, if the Employer's Requirements state that obscured glass is required in certain windows, the employer is certainly entitled to ensure that the contractor has provided such glass in the documents. However, if the employer simply would have constructed a detail in a different way from the contractor's own perfectly adequate detail, the detail cannot be rejected on that basis.

Any documents not returned within 14 days are to be regarded as marked 'A', and documents marked 'B' or 'C' must include written reasons why they are not in accordance with the contract. The explanation of the marking is as follows:

A The contractor must carry out the Works strictly as the submitted document.
B The contractor may carry out the Works as the submitted document if the employer's comments are incorporated and an amendment copy of the document is submitted to the employer promptly. This is presumably intended to deal with minor infringements where the document differs from the contract documents to a minor degree.
C The contractor must take account of the employer's comments and resubmit for comment as amended as soon as it reasonably can do so; alternatively the contractor may disagree with the employer and then must follow the procedure in paragraph 7.

Paragraph 6 states the obvious, in that the contractor must not carry out any work in accordance with documents marked 'C', and emphasises that the employer has no liability to pay for any work carried out which is not shown on documents marked either 'A' or 'B'. What is clear is that the contractor must submit documents for the whole of the Works. There is no question of the contractor constructing any part of the Works on the basis of it being obvious what is to be done. The employer can simply refuse to pay. This useful power is hidden away in the depths of this schedule. It should be highlighted by the users of this contract.

If the contractor does not agree with any comment made about a submitted document by the employer, it has 7 days from receipt of the comment to write to the employer. The written notice should inform the employer that the contractor considers that to comply with the comment would result in a change, stating why the contractor is of that view. The employer then has 7 days from receipt of the notice to confirm or withdraw the comment in question. If the employer confirms the comment, the contractor must amend the document and resubmit it, although no timescale is indicated. It is easy to see how this procedure works in relation to a document marked 'C', but because the contractor might disagree with a comment written by the employer about a document marked 'B', it is clear that where there is such disagreement, the contractor must resubmit the amended document and await further marking by the employer before proceeding to carry out the Works shown on that document.

Paragraph 8.1 stipulates that the employer's confirmation or withdrawal of the comment in response to the contractor's notice does not indicate that the employer has accepted that the document, amended or not, is in accordance with the contract. Neither does it indicate that if the contractor had complied with the comment, it would have resulted in a change. In other words, the contractor is not to conclude that confirmation or withdrawal of an employer's comment in any way results in a change. That does not mean that the contractor cannot subsequently successfully maintain that a confirmed comment will result in a change under the provisions of clause 5. It will be a matter of fact. However, the contractor must beware of paragraph 8.2, which specifically states that a comment, about which the contractor has not given notice to the employer under paragraph 7, will not result in a change. The moral is clear. If the contractor believes that compliance with one of the employer's comments will result in a change, it must give notice under paragraph 7; otherwise, even if the comment does qualify as a change under clause 5, it will not be so treated and the contractor will have lost its chance for payment.

4.4.2 Copyright, royalties and patents

The contractor must include for all royalty payments etc. which are payable in relation to any supply or use of anything in connection with the Works (clauses 2.19 and 2.20). This will generally include everything expressed or inferred in the Requirements or the Proposals. In addition, the contractor indemnifies the employer against all claims which may be brought against him as a result of any infringement by the contractor of any patent rights or the like. The effect of this provision is that the contractor agrees to reimburse the employer all costs in connection with such infringement without limitation.

If the contractor infringes any rights as a result of complying with the employer's instructions, any money which the contractor is liable to pay will be added to the contract sum as reimbursement to the contractor.

4.4.3 Access to the Works

Under clause 3.1, the contractor is obliged to give access to the Works, and other places where work is being prepared, for the employer's agent and any person authorised by the employer. The contractor must also include terms in its sub-contracts to achieve similar rights of access to sub-contractors' workshops. The contractor must do everything reasonably necessary to make such rights effective. There is an important proviso that the contractor and any sub-contractor may impose reasonable restrictions to safeguard their proprietary rights.

Chapter 5
The Employer's Powers and Duties

5.1 *Employer's agent*

The employer's agent is provided for in article 3. The choice of the term is deliberate and there is no sense in which the employer's agent is performing the same function as an architect under SBC. Under the traditional form of contract, the architect not only acts in an agency capacity but also owes a duty to the employer to act fairly between the parties: *London Borough of Merton* v. *Stanley Hugh Leach Ltd* (1985). The employer's agent, although possibly an architect, is generally thought to have no such duty under DB, although this is arguably not so. The Court of Appeal in *Balfour Beatty Civil Engineering Ltd* v. *Docklands Light Railway Ltd* (1996) said in relation to a different form of contract:

> 'We would . . . have wished to consider whether an employer vested with the power . . . to rule on his own and a contractor's rights and obligations, was not subject to a duty of good faith substantially more demanding than that customarily recognised in English contract law.'

The decision in this case has subsequently been questioned in *Beaufort Developments (NI) Limited* v. *Gilbert Ash NI Limited* (1998), but not on this point. There is little doubt, however, that the employer's agent under this contract does not have the same status in the eyes of a court or an arbitrator as does an independent architect engaged under a traditional contract: *J F Finnegan* v. *Ford Seller Morris Developments (No.1)* (1991).

At first sight, the last sentence of clause 2.1.4 appears to invest the employer with considerable powers. It states that the contractor is bound by any decision of the employer 'made under or pursuant to these Conditions'. However, the end of the sentence makes clear that the contractor is bound only to the extent that it does not challenge it in adjudication, arbitration or litigation, as the case may be. This perfectly sensible provision requires that when the employer has acted when obliged or empowered to do so in accordance with the contract, that kind of decision is binding. This is an altogether different concept to the provision in some contracts that the architect/engineer/contract administrator's decision is final and binding. That kind of provision would not be open to review in the dispute resolution procedure.

The principal (the employer) is bound by the properly authorised actions of the agent. It is important to establish the extent of the agent's authority. It may be actual or apparent (ostensible). The employer's agent's actual authority is defined by the terms of the agreement with the employer. The apparent authority of the employer's agent may be quite different. Apparent authority is the authority the agent seems to possess as viewed by persons other than the employer. The agent's authority, so far as the contractor is concerned, will be laid down in the terms of the contract DB and in the Employer's Requirements. Therefore, if the authority given to the agent by DB is not matched by the agent's actual authority under the agreement, the contractor is entitled to take account of the apparent authority and the employer will be responsible for the agent's actions pursuant to the provisions of DB. The agent, however, having exceeded his or her actual authority, will be liable to the employer for the consequences. It is, therefore, crucially important that the employer's agent ensures that the agreement with the employer gives at least the same degree of authority as apportioned in DB and the Employer's Requirements, and preferably rather more. In addition, the exact scope of that authority must be clearly defined.

An agent's duties to the principal are as follows:

- To act. The employer may sue if the agent fails to act at the appropriate time.
- To obey instructions. The proviso is that the instructions must be lawful and also reasonable.
- To declare to the principal if there is any conflict of interest.
- To keep proper accounts.
- Not to take any secret bribe or profit. Breach of this duty entitles the principal to recover damages which may include the amount of the bribe or profit.
- Not to delegate without authority. This is important in the context of the duties which DB lays at the agent's door.

In the circumstances envisaged by DB, the agency relationship will almost certainly be the subject of an express appointment. It should be in writing to reduce the possibilities of misunderstandings. There are, however, three other ways in which agency may arise, which are included here for the sake of completeness. If the employer acts to others as though a person is the agent and that person is acting like an agent, agency may arise by implication. If a person acts for another in an emergency, the agency may be created by necessity. Finally, it sometimes happens that a person performs some action for another and the other then ratifies it. Ratification requires two conditions to be satisfied: the agent must perform the action on behalf of the principal and the principal must have been capable of carrying out the action at the time it was performed.

Article 3 refers to the employer's agent as a 'person'. Although the word is not defined at all in this contract, in normal principles of interpretation the term may be taken to mean more than an individual. Reference to the Interpretation Act 1978 suggests that it can be taken to include a person, or a body corporate or incorporate. Indeed it was defined in a similar way in the JCT 98 contract where 'partnership' was expressly included. The person nominated as agent may well be an employee of the employer, although there appears to be no reason why the name of a company cannot be inserted. This is likely to be the case if the employer is a large corporation with its own in-house professional staff. In other cases, the agent will probably be an appropriate professional person such as an architect, an engineer or a surveyor. Where the person nominated is the head of an in-house department, it will be accepted that the duties of agent will be carried out by members of staff, and the same is true where a private consultant is named. Article 3 clearly provides for the employer to nominate a replacement without any reason being given and there appear to be no formalities involved (other than the obvious necessity of written notice) and no right for the contractor to raise any objection. The person nominated as agent need not necessarily be a professional, nor need be involved in construction. Clearly, however, the employer would be well advised to appoint a construction professional in this role.

The employer will often appoint an architect in the first instance to assist in preparing the Employer's Requirements and any drawings required before the contract is executed, and to give general advice. It makes a great deal of sense to appoint that person as employer's agent unless, of course, a consultant switch or novation is to be operated (see section 3.4). Whoever is appointed as agent under this contract should be quite clear that the role is far removed from the role of an architect in a traditional contract. The reality is that many architects in this position proceed as they would under a traditional contract, issuing certificates and instructions, ascertaining loss and expense, and so on. What is required is a completely different approach, which should become clear in this chapter. There appears to be no reason why the employer should not have other advisors, such as quantity surveyors and engineers.

The employer's agent is expressly referred to only in clauses 2.7.3 and 3.1 in relation to the availability of drawings and access to the site respectively. The employer's agent or any other person authorised by the employer or the agent is to have access at all reasonable times to the Works, workshops and other places where work is being prepared for the contract. The contractor is also to obtain similar access rights from sub-contractors. The contractor and any sub-contractor may impose whatever restrictions are necessary to safeguard their trade secrets. This clause is very similar to the equivalent clause in SBC.

The contract empowers the agent to act for the employer in receiving or issuing applications, consents, instructions, notices, requests or statements or to otherwise act for the employer under any of the conditions. The article could scarcely be more broadly drafted. An employer who wishes some other arrangement to apply must give written notice to the contractor to that effect. It is possible that the employer might wish to reserve some decisions or to specify that another person shall act in relation to particular clauses. For example, it might be specified that a quantity surveyor acts for the employer in every clause which has a particular cost aspect. Clauses 4 and 5 are obvious candidates for this treatment. There is no reason why article 3 should not be redrafted to allow the employer to appoint a number of persons as employer's agents for particular purposes, as set out in the Employer's Requirements. To avoid confusion, it is essential that the respective powers and duties are clearly set down. For a list of the employer's/employer's agent's powers and duties, see Figs 5.1 and 5.2.

5.2 *Express and implied terms*

Like the contractor, the employer is governed by express and implied terms. In most instances, an implied term will be excluded or modified by express terms of the contract. However, there are two terms of great importance concerning the employer which the law will imply into every building contract irrespective of its express terms. These terms are fundamental to the proper carrying out of the contract.

Clause	Power	Comment
2.2.1	Consent to the substitution by the contractor of anything in the Employer's Requirements, the Contractor's Proposals or the documents referred to in clause 2.8.	Consent must be in writing and must not be unreasonably delayed or withheld. The consent does not relieve the contractor of its other obligations.
2.2.4	Request the contractor to provide reasonable proof that materials and goods comply with clause 2.2.	
2.4	Defer giving possession for not more than 6 weeks.	Where clause 2.4 is stated in the contract particulars to apply.
2.5.1	Use or occupy the site of the Works before the issue of practical completion certificate.	With contractor's written consent
2.6	Have work not forming part of the contract carried out by the employer or the employer's persons.	If the Employer's Requirements so provide and give the contractor the necessary information. If not, the contractor's consent is required under clause 2.6.2.
2.14.1	Decide between discrepant items or accept the contractor's proposed amendment.	After contractor has notified employer of proposed amendment to remove discrepancy within Contractor's Proposals.
2.14.2	Agree contractor's proposed amendment or decide how discrepancy must be dealt with.	After contractor has notified employer of proposed amendment to deal with discrepancy within Employer's Requirements not dealt with by Contractor's Proposals.
2.15.1	Consent to any amendment proposed by the contractor for removing the divergence and note the amendment on the specified documents.	Consent must not be unreasonably withheld.
2.21	Consent in writing to the removal of unfixed materials and goods delivered to, placed on or next to the Works.	Consent must not be unreasonably withheld.
2.25.4	Fix an earlier completion date than that previously fixed under clause 2.25.1 or pre-agree adjustment if it is fair and reasonable to do so in the light of any subsequently issued omission instructions.	This can only be done after an extension of time under clause 2.25.1 or pre-agreed adjustment.
2.25.5	Review extension of time granted and fix a new completion date or confirm. Fix an earlier date for completion.	May be carried out at any time after completion date if this occurs before existing practical completion, but no later than 12 weeks after practical completion. Having regard to instructions requiring an omission issued after the last extension of time.

Fig. 5.1 Employer's powers.

2.29.2	Give written notice requiring the contractor to pay liquidated damages *or* give written notice of intention to deduct liquidated damages.	The employer's notices under clause 2.28 and 2.29.1.2 must have been given.
2.3	Take partial possession of the Works before practical completion.	With the contractor's consent.
2.35	Issue instructions that defects etc. be not made good.	An appropriate deduction must then be made from the contract sum.
2.35.2	Issue instructions that defects etc. be made good.	No such instructions can be issued after delivery of the schedule of defects or after 14 days from the expiry of the defects liability period.
2.37	Reasonably require information related to the contractor's design documents.	Showing the work as-built.
2.38.2	Copy and use the contractor's design documents for any purpose.	Subject to all monies due and payable having been paid.
3.1	Have access to the work and to the workshops and other places of the contractor where work is being prepared for the contract.	This refers to the employer's agent and any other person authorised by the employer or the employer's agent. The right is exercisable at reasonable times, that is during normal working hours.
3.3	Consent in writing to sub-letting all or part of the Works, including design.	Consent must not be unreasonably withheld or delayed.
3.5	Issue instructions to the contractor on any matter in respect of which the employer is empowered to do so by the contract.	The contractor must immediately comply with such instructions *except* one requiring a change within clause 5.1.2, in which case the contractor has the right of reasonable objection.
3.6	Issue a written notice to the contractor requiring compliance with an instruction. Employ and pay others to give effect to the instruction. Recover all costs in connection therewith by deduction from the contract sum.	If the contractor fails to comply with the 7-day compliance notice.
3.7.3	Confirm in writing any non-written instruction which has been given.	Within 7 days of its issue.
3.7.4	Confirm any oral instruction in writing at any time before the final account and final statement become conclusive as to the balance due under the contract.	This is a long-stop provision.
3.8	Invoke the dispute resolution procedures.	If the employer wishes a decision as to whether the specified clause provision in fact authorises the instruction. The request must be made before the contractor complies with the instruction.

Fig. 5.1 *Continued*

3.9.1	Issue instructions effecting a change in the Employer's Requirements	A change is a variation. No change may be effected (unless the contractor consents) which is or which will make necessary an alteration or modification in the design element of the Works.
3.10	Issue instructions regarding the postponement of any work to be executed under the contract.	
3.12	Issue instructions requiring the contractor to open up for inspection, testing, etc. of work, goods and materials.	The cost will be added to the contract sum unless the tests etc. prove adverse to the contractor or where such costs have been provided for in the Employer's Requirements or Contractor's Proposals.
3.13.1	Issue instructions requiring removal from site of any work, materials or goods not in accordance with the contract.	
3.13.2	Issue reasonably necessary instructions re change following clause 3.13.1 instruction.	After consulting the contractor. No addition to the contract sum is to be made and no extension of time given.
3.13.3	Issue reasonable instructions to open up work or carry out tests to establish to employer's satisfaction the likelihood of further similar non-compliance.	After having due regard to the code of practice appended to the conditions.
3.14	Issue instructions re change reasonably necessary.	After failure to comply with clause 2.1. After consulting contractor no addition is to be made to the contract sum and no extension of time is to be given.
4.10.2	Exercise any right to withhold from monies due against any amount otherwise due.	Whether or not retention is included in such amount.
4.10.4	Give written notice to the contractor specifying amount proposed to be withheld or deducted from the due amount, the grounds and amount of withholding attributable to each of the grounds.	Not later than 5 days before the final date for payment.
4.12.5	Give written notice to the contractor 3 months after practical completion. Prepare or have prepared a final account and statement.	If the contractor does not submit final account and statement within 2 months of the date of the notice. If the contractor does not so submit.
4.12.9	Give written notice to the contractor specifying amount proposed to be withheld or deducted from the due amount, the grounds and amount of withholding attributable to each of the grounds.	Not later than 5 days before the final date for payment.

Fig. 5.1 *Continued*

4.16	Deduct and retain the retention percentage, holding the same as trustee.	
4.19.2	Reasonably require information and details.	If the contractor makes application for loss and/or expense.
6.4.2	Require the contractor to produce documentary evidence that clause 6.4.1 insurances are properly maintained.	The power must not be exercised unreasonably or vexatiously.
6.4.3	Effect the necessary insurance if the contractor has failed to insure or continue to insure.	The premiums may be deducted from monies due or to become due to the contractor or can be recovered as a debt.
6.5.2	Approve insurers.	In regard to insurance against damage to property other than the Works – employer's liability.
6.5.3	Insure against damage to property other than the Works – employer's liability.	If the contractor fails to insure when so stated in the Employer's Requirements.
6.11.3	Reasonably request documentary evidence.	That professional indemnity insurance is being maintained.
6.15.2	Employ other persons to carry out remedial measures and an appropriate deduction is to be made from the contract sum.	If the contractor fails to do so.
7.1	Assign the contract.	Only if the contractor consents in writing.
7.2	Assign to any transferee or lessee the right to bring proceedings in the employer's name or enforce any contractual terms made for the employer's benefit. The assignee is estopped from disputing enforceable agreements reached between employer and contractor related to the contract.	If the employer transfers leasehold or freehold interest or grants a leasehold. If made prior to the date of the assignment.
7B.1	By notice to the contractor convey rights to the funder identified in the notice.	Where clause 7B is stated in the contract particulars to apply.
7C	By notice to the contractor identifying the purchaser/tenant and interest in the Works, require a warranty.	Where clause 7C is stated in the contract particulars to apply. Within 14 days.
7D	By notice to the contractor require warranty for funder.	Where clause 7D is stated in the contract particulars to apply. Within 14 days.
7E	Approve amendments in warranty for purchaser, tenant, funder or employer.	Proposed by a sub-contractor. Approval not to be unreasonably delayed or withheld.

Fig. 5.1 *Continued*

8.4.1	Serve written notice on the contractor by actual, special or recorded delivery specifying the default.	If the contractor: • wholly or substantially suspends design or construction of the Works without reasonable cause; *or* • fails to proceed regularly and diligently; *or* • does not comply with a written notice to remove defective work; *or* • fails to comply with the assignment or sub-contracting clause; *or* • fails to comply under the contract with the CDM Regulations.
8.4.2	Terminate the contractor's employment by written notice served by actual, special or recorded delivery.	The notice must not be given unreasonably or vexatiously. It can be served only if the contractor continues its default for 14 days after receipt of the default notice. The notice must be served within 10 days of the expiry of the 14 days.
8.4.3	Terminate the contractor's employment by written notice served by actual, special or recorded delivery.	If the contractor ends the default or the employer does not terminate and the contractor thereafter repeats the default.
8.5.1	Terminate the contractor's employment.	If the contractor is insolvent.
8.5.3.3	Take reasonable measures to ensure that site materials, the site and the Works are adequately protected.	From the date when the contractor became insolvent.
8.6	Terminate the contractor's employment under this or any other contract by written notice served by recorded or special delivery.	Where the contractor is guilty of corrupt practices.
8.7.1	Employ and pay others to carry out and complete the design and construction of the Works and make good defects, take possession of the site and make use of the contractor's equipment etc.	After termination under clauses 8.4–8.6.
8.7.2.1	Require the contractor to remove from the Works its temporary buildings etc.	
8.7.2.3	Require the contractor to assign to the employer the benefit of any sub-contracts etc.	Within 14 days of termination unless termination was due to certain types of insolvency.
8.8.1	Decide not to have the Works carried out and completed.	The employer must notify the contractor within 6 months of termination.

Fig. 5.1 *Continued*

8.11	Give 7 days' written notice of termination.	If the carrying out of substantially the Works is suspended for a period named in the contract particulars by reason of: • *force majeure*; *or* • employer's instructions resulting from local authority or statutory undertaker's negligence; *or* • loss by specified perils; *or* • civil commotion, terrorist activity; *or* • exercise by UK government of statutory power; *or* • delay in receipt of development control permission.
	By further notice terminate the contractor's employment.	Failing cessation within the 7-day period.
9.1	Seek to resolve dispute by mediation.	By agreement.
Schedule 2, para. 1.2	Consent to the appointment of manager.	Must be in writing. For this clause to apply, para. 1 must be stated to apply in the contract particulars. It replaces clause 3.2.
	Consent to replacement of manager.	Must be in writing and not be unreasonably withheld or delayed.
Schedule 2, para. 1.3	Request the manager and any of the contractor's servants, sub-contractors, etc. to attend meetings.	Must be reasonable.
Schedule 2, para. 1.4	Inspect the manager's records.	
Schedule 2, para. 2.1.3	Consent to a person selected.	Following para. 2.1.2.2 change requiring the contractor to select another person to carry out named sub-contract work.
Schedule 2, para. 2.1.5	Consent to the termination of named sub-contractor's employment.	
Schedule 2, para. 4.2.5	Either: • instruct compliance and para. 4 shall not apply; *or* • withdraw the instruction.	If the parties cannot agree within 10 days of receipt of estimates following an instruction.
Schedule 2, para. 5.4	Request such information as may reasonably be required in support of the contractor's estimate	Within 21 days of receipt of an estimate of loss and/or expense under para. 5.2 or 5.3.

Fig. 5.1 *Continued*

Schedule 3, para. A.2	Accept deposit of policies and premium receipts. Insure against all risks and deduct sums from monies due or recover them as a debt.	In regard to insurance against all risks to be taken out by the contractor. If the contractor has failed to insure or continue to insure.
Schedule 3, para. A.3	Inspect documentary evidence or the policy.	If the contractor maintains a policy independently of its obligations under the contract and it is in joint names.
Schedule 3, para. A.5.2	Instruct contractor not to renew terrorism cover.	If employer is a local authority. In lieu of adjustment of contract sum under para. A.5.1.
Schedule 3, para. C.4.4	Terminate the contractor's employment under the contract. Invoke the dispute resolution procedures.	If the Works are damaged by para. C.2 risks and it is just and equitable to do so. Within 7 days of the contractor serving notice terminating its employment.

Fig. 5.1 *Continued*

Clause	Duty	Comment
2.3	Give possession of the site to the contractor on the date for possession.	
2.5	Notify insurers under insurance options A, B or C.2 and obtain confirmation that use or occupation will not prejudice insurance.	Or the contractor may notify.
2.7.1	Be custodian of the Employer's Requirements, Contractor's Proposals and the contract sum analysis.	These documents must be made available for the contractor's inspection at all reasonable times.
2.7.2	Provide the contractor with a copy of the contract, the Employer's Requirements, Contractor's Proposals and contract sum analysis certified on behalf of the employer.	This must be done immediately after the contract is executed unless the contractor has previously been provided with the documents
2.7.4	Not to divulge to third parties or use any of the specified documents supplied by the contractor other than for the purposes of the contract.	
2.9	Define the boundaries of the site.	
2.10.1	Issue instructions on divergencies between the Employer's Requirements and the definition of the site boundary. The instruction is deemed to be a change and is valued accordingly.	

Fig. 5.2 Employer's duties.

2.10.2	Give written notice to the contractor specifying the divergence.	If such a divergence is found.
2.13	Issue instructions.	If contractor finds and notifies any discrepancy.
2.14.1	Decide between discrepant items or otherwise accept contractor's amendments as proposed.	Where there is a discrepancy in the Contractor's Proposals and the contractor has informed the employer of it.
2.14.2	Agree the proposed amendment or decide how the discrepancy is to be dealt with.	Where there is a discrepancy in the Employer's Requirements not addressed by the Contractor's Proposals.
2.15.1	Immediately give written notice to the contractor of any divergence between statutory requirements and either the Employer's Requirements or the Contractor's Proposals. Note the agreed amendment on the contract documents.	The Employer's Requirements include any change (clause 5) and the duty arises if such a discrepancy is found.
2.15.2.3	Issue an instruction requiring a change.	If amendment to Employer's Requirements becomes necessary to comply with statutory requirements.
2.25–2.25.3	Make in writing a fair and reasonable extension of time for completion by fixing a later date as the completion date and stating (a) which of the relevant events have been taken into account and the extension of time allotted to each and (b) the extent to which reduction in time has been given to any omission instruction issued since the fixing of the previous completion date. Notify contractor in writing if it is not fair and reasonable to fix a later date.	It must become apparent that the progress of the Works is being or is likely to be delayed; *and* the contractor must give written notice of the cause of the delay and supply supporting particulars and estimate; *and* the reasons for the delay must fall within the list of relevant events. The decision must be given not later than 12 weeks from receipt of particulars etc. If the current completion date is less than 12 weeks away the employer must endeavour to give the decision before such date.
2.25.5	Write to the contractor either: • fixing a later completion date than that previously fixed if it is fair and reasonable to do so having regard to the relevant events; *or* • fixing an earlier completion date, likewise in the light of any omission instructions issued subsequently; *or* • confirming the completion date previously fixed.	It is a duty to review the situation whether the relevant event has been notified or not. No decision under clause 2.25.4 or 2.25.5.2 can fix a date earlier than the date for completion stated in the contract.

Fig. 5.2 *Continued*

2.27	Give the contractor a written statement of practical completion.	When the Works or a section have reached practical completion and the contractor has complied sufficiently with clauses 2.37 and 3.18.4.
2.28	Issue written notice to the contractor stating that it has failed to complete the construction of the Works by the completion date. Issue further notice as necessary.	 After a new completion date has been fixed after the first issue of a clause 2.28 notice.
2.29.3	Pay or repay liquidated damages to the contractor.	Where a later completion date is fixed.
2.32	Issue a notice when all defects have been made good.	In the relevant part.
2.35.1	Deliver to the contractor a schedule of defects which appear within the rectification period.	The defects specified must be due to the contractor's failure to comply with its contractual obligations. The schedule of defects must be issued not later than 14 days after the rectification period expires.
2.36	Issue a notice that the contractor has made good all defects in the schedule.	Once the contractor has discharged its liability. The notice must not be unreasonably withheld or delayed.
3.7.1	Issue all instructions in writing.	There is a procedure for the confirmation of oral instructions.
3.8	Specify in writing the contract clause empowering the issue of an instruction.	On the contractor's written request.
3.9.4	Vary the terms of the instruction to remove the contractor's objection.	If the contractor objects to an instruction under its obligations under the CDM Regulations.
3.11	Issue instructions to the contractor on the expenditure of provisional sums included in the Employer's Requirements.	
3.16	Issue instructions on antiquities found.	If the contractor reports a find of antiquities etc. under clause 3.15.
3.18	Comply with the CDM Regulations.	
3.18.1	Ensure that the CDM co-ordinator carries out all its duties under the CDM Regulations and that the principal contractor carries out all its duties under the CDM Regulations.	Where the contractor is not the CDM co-ordinator. Where the contractor is not the principal contractor.

Fig. 5.2 *Continued*

3.19	Immediately notify the contractor in writing of the name and address of the new CDM co-ordinator or the new principal contractor.	If the employer replaces the existing CDM co-ordinator or the principal contractor.
4.6	Pay the advance payment.	If clause 4.6 applies and provided that any bond required has been given. This clause does not apply if the employer is a local authority.
4.7	Make interim payments to the contractor.	In accordance with clause 4 and alternative A or B as specified in the contract particulars.
4.10.3	Give written notice specifying the amount of proposed payment, the basis of calculation and to what it relates.	Not later than 5 days after receiving application.
4.10.5	Pay the amount stated in the clause 4.10.3 notice.	Subject to any clause 4.10.4 notice. If no clause 4.10.3 notice, the amount payable is to be calculated in accordance with clause 4.8.
4.10.6	Pay the contractor simple interest.	If the employer fails to pay the amount due by the final date for payment. Payment is treated as a debt and the rate is 5% over Bank of England base rate.
4.12.8	Give a written notice to the contractor specifying the amount of payment proposed in respect of any balance due to the contractor in the final statement or in the employer's final statement.	Not later than 5 days after the final statement becomes conclusive as to the balance due.
4.12.10	Pay the balance stated as due in the final statement.	Subject to any clause 4.12.9 notice. If the employer does not give a clause 4.12.8 notice.
4.12.11	Pay the contractor simple interest.	If the employer fails to pay the amount due by the final date for payment. Payment is treated as a debt and the rate is 5% over Bank of England base rate.
4.16.2	Place the retention in a separate designated banking account and inform the contractor in writing that the amount has been so placed.	Unless the employer is a local authority, if the contractor so requests. The retention is to be banked at the date of each interim payment. The employer gets the interest.

Fig. 5.2 *Continued*

4.19	Reimburse the contractor for any direct loss and/or expense caused by matters affecting regular progress of the Works.	If the contractor makes written application within a reasonable time of it becoming apparent, and the necessary procedural and other conditions of the clause have been satisfied. It is implied that an ascertainment will be made as under SBC – which should be dealt with in the Employer's Requirements.
6.9	Ensure that the joint names policies referred to in paragraphs B.1 or C.2 of schedule 3 either: • provide for recognition of each sub-contractor as an insured; *or* • include insurer's waiver of rights of subrogation.	If option B or C applies. Also applies to any joint names policy taken out by the employer under paragraph A.2.
6.10.2	Give written notice to the contractor before the cessation date: that the Works continue; *or* that the contractor's employment must terminate.	After receipt of notification from insurers.
6.14	Comply with the Joint Fire Code and ensure such compliance by the employer's persons.	
6.15.1	Copy notice to contractor.	If insurers notify the employer of required remedial measures.
8.7.4	Set out an account in a statement.	Within a reasonable time of completion of the Works and making good defects by others.
8.8.1	Notify the contractor in writing within 6 months of the date of termination. Send a written statement of account to the contractor.	If the employer decides not to have the Works completed. If the employer so notifies.
8.10.2	Immediately inform the contractor in writing if the employer makes a proposal, gives notice of a meeting or becomes the subject of any proceedings or appointment in clause 8.1.	
8.12.3	Prepare an account.	With reasonable dispatch if the employer opts to prepare it and if the contractor has discharged its obligation to provide documents within 2 months.
8.12.5	Pay the contractor the amount properly due.	After taking amounts previously paid into account. Payment must be made within 28 days of submission.
9.4	Serve on the employer a notice of arbitration.	If the employer wants a dispute referred to arbitration.

Fig. 5.2 *Continued*

Schedule 1, para. 2	Return one copy marked 'A', 'B' or 'C' within 14 days of submission or expiry of period for submission.	If contractor submits contractor's design documents.
Schedule 1, para. 4	Identify in writing why it is not in accordance with the contract.	If the employer marks a contractor's design document 'B' or 'C'.
Schedule 1, para. 7	Confirm or withdraw comment.	Within 7 days of receipt of notification from contractor that compliance would give rise to a change.
Schedule 2, para. 2.1.2	Either: • remove the reason for inability; *or* • omit the named sub-contract work from the Employer's Requirements and issue instructions re the execution of such work.	If the contractor is unable to enter into a sub-contract with the named person for a *bona fide* reason.
Schedule 2, para. 3.2	Correct errors in description or quantity in bills of quantity.	The correction is to be treated as a change in the Employer's Requirements.
Schedule 2, para. 3.4	Provide amplification of any bills of quantities included in the Employer's Requirements.	Where fluctuations option C applies.
Schedule 2, para. 5.4	Give the contractor written notice either: • that the employer accepts the estimate; *or* • that the employer wishes to negotiate and in default of agreement to refer the issue to the adjudicator; *or* • that clause 4.19 shall apply.	Within 21 days from receipt of paragraph 5.2 or 5.3 estimate.
Schedule 3, para. A.4.4	Pay insurance monies received to the contractor.	By instalments under clause 4.14 alternative B even if alternative A is applicable to other payments.
Schedule 3, para. B.1	Maintain proper insurances against all risks.	Where the employer has undertaken the risk in the case of new Works.
Schedule 3, para. B.2.1	Produce receipts etc. to the contractor at its request.	Unless the employer is a local authority.
Schedule 3, para. B.2.2	Produce copy of cover certificate from insurer that terrorism cover is being provided.	If the employer is a local authority.
Schedule 3, para. C.1	Maintain adequate insurances against specified perils.	In the case of existing structures.
Schedule 3, para. C.2	Maintain insurance against all risks for work of alterations or extensions.	In joint names.
Schedule 3, para. C.3.1	Produce insurance receipts etc.	If the contractor so requests unless the employer is a local authority.
Schedule 3, para. C.3.2	Produce copies of cover certificates from insurers that terrorism cover is provided under each policy.	If the employer is a local authority.

Fig. 5.2 *Continued*

The two terms are complementary and they may be expressed in the context of this contract as follows:

- The employer and the employer's agent will do all that is reasonably necessary to enable the contractor to carry out and complete the Works in accordance with the contract: *Luxor (Eastbourne) Ltd* v. *Cooper* (1941).
- Neither the employer nor the employer's agent must hinder or prevent the contractor from carrying out and completing the Works in accordance with the contract: *William Cory & Sons* v. *City of London Corporation* (1951).

These terms are capable of very broad interpretation and often form the basis of substantial claims by the contractor. For example, on one hand the employer must make sure to give the contractor all necessary decisions in good time, and to ensure the site is available and that there is a good access; on the other hand, the access must not be closed, decisions must not be refused and the work must not be stopped. If the employer fails to comply with these two implied terms, the contractor's obligation to complete by the contract completion date is removed and its duty is simply to complete the Works in a reasonable time. The problems this would cause are largely avoided in standard building contracts by the inclusion of a clause allowing extension of the period for carrying out the work for most of such failures (see Chapter 8).

5.3 *Instructions*

5.3.1 Procedure

The issue of instructions is covered by clause 3.5 which closely resembles its counterpart in SBC. All instructions issued by the employer or by the employer's agent must be in writing. The contract does not prescribe any special form for the purpose. It is sufficient if the words are presented to the contractor in permanent visible form, and provided it is clear that the words are instructing the contractor to do something. Most instructions will be in the form of a letter, but they can be written on a pad on site, on the back of an envelope or on the side of a brick. The more bizarre methods are not advocated. An architect acting as agent may well use a standard form for issuing instructions, but care must be taken to strike out the words 'Architect's Instruction' and substitute 'Employer's Agent'. Whoever issues instructions will find it helpful to use standard forms for the purpose because they make the checking process much easier at the end of the project.

After stating that all instructions must be in writing, the contract proceeds at some length to set out the procedure if an instruction is issued 'otherwise than in writing', i.e. orally (clause 3.7). Oral instructions are of no immediate effect and if the contractor complies with such an instruction, it does so at its own risk. Depending on the content of the instruction, the contractor may be in breach of its obligation to carry out the work in accordance with the contract documents. For example, if the employer gives an oral instruction requiring the enlargement of a restaurant terrace to seat 100 rather than 50 people and the instruction is not confirmed but the contractor complies, it will result in work which is not in accordance with the contract. In theory, the employer can order rectification or removal under clause 3.13.1, but in *G Bilton & Sons* v. *Mason* (1957) it was held that a contractor's compliance with unconfirmed architect's instructions was a good defence to a claim for breach of contract. Evidence would have to be brought that the oral instruction was given, of course. Under clause 3.7.2, after receiving an oral instruction, the contractor has 7 days in which to write to the employer to confirm it. The employer then has a further 7 days in which to dissent. If the employer does not dissent, the instruction takes effect, not from the date it was issued but from the expiry of the employer's 7-day dissent period.

If the employer confirms an oral instruction within 7 days under clause 3.7.3, the instruction takes effect from the date of confirmation and the contractor's duty to confirm is removed. If neither contractor nor employer confirms an oral instruction in writing but the contractor has nevertheless complied, clause 3.7.4 provides that the employer may confirm it in writing at any time up to the date at which the final account and final statement become conclusive. Strangely, in this instance, the instruction is then deemed to have been issued on the date that the oral instruction was given. This is intended as a safeguard for the contractor and to provide a mechanism whereby the contractor can secure payment if it complies with an oral instruction. A contractor who leaves the confirmation of an oral instruction to this late date, however, is asking for trouble. With the best will in the world, which may not be much in evidence, the memory of the employer's agent will dim and the agent may even be replaced with another. There is generally little excuse for oral instructions. They are a sign of laziness – probably on both sides. It is good practice for the contractor to keep a duplicate book on site for the benefit of the employer's agent. Oral instructions can be jotted down, signed and dated, and there is no need for delay or complex confirmations. In these days of the fax machine, the days of the telephoned instruction should be at an end.

In theory, there should be very few instructions issued by or on behalf of the employer under this form of contract, because the issue of many instructions removes much of the risk which the contractor otherwise takes in respect of the date for completion and the price. The reality can be different, perhaps because the employer has not properly finalised the requirements in the contract documents or perhaps because the employer does not appreciate the crucial differences between this and other forms of procurement (see Chapter 1).

Clause 3.5 provides that the contractor must forthwith comply with all instructions issued by the employer. The meaning of 'forthwith' is that the contractor must comply as soon as it reasonably can: *London Borough of Hillingdon* v. *Cutler* (1967). There are a number of important provisos:

- The employer must be expressly empowered under the contract to issue the instruction in question. A list of instructions empowered by the contract is given in Fig. 5.3. Clause 3.8 gives the contractor power to request the employer to specify in writing the empowering clause in the contract. The employer must comply forthwith and if the contractor then complies with the instruction, it will be deemed 'duly given' under the clause specified by the employer. The most important reason for this power is probably the valuation of change instructions, clauses 5.2–5.7. The effect of this clause is that even if the employer is wrong in believing that the instruction is empowered under the specified clause,

Clause	Instruction
2.10.1	In regard to discrepancy between Employer's Requirements and definition of site boundary.
2.15.2.3	After amendment to Employer's Requirements, to which clause 2.1.2 refers, becomes necessary to comply with statutory requirements.
2.35.1	Schedule of defects.
2.35.2	Requiring defects etc. to be made good.
3.5	Employer's instructions in general.
3.9.1	Effecting a change in the Employer's Requirements.
3.10	Postponing work.
3.11	Expenditure of provisional sums.
3.12	To open up for inspection or carry out testing.
3.13.1	Removal from site of work, materials or goods.
3.13.2	Reasonably necessary change after defective work.
3.13.3	Reasonably necessary change to establish likelihood of further non-compliance.
3.14	Reasonably necessary change after failure to comply with clause 2.1.
3.16	Regarding antiquities, including excavation, examination or removal by third parties.
Schedule 2, para. 2.2.2	Omitting named sub-contract work and regarding the execution of that work.
Schedule 2, para. 4.5.1	Compliance with instruction.

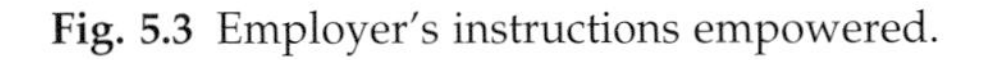

Fig. 5.3 Employer's instructions empowered.

the contractor is entitled to whatever benefits would flow from a properly empowered instruction if it is queried before compliance. As an alternative, the contractor or indeed the employer may invoke the relevant procedures for the resolution of disputes to decide whether the instruction is empowered by the specified clause. The relevant procedure usually will be adjudication unless both parties wish the matter to be referred immediately to arbitration or dealt with in litigation (whichever method is included in the contract). It is thought that the contractor has the right to await the outcome before complying. Sensibly, the contractor may decide to comply pending the result of the dispute procedure if it can get written agreement from the employer that such compliance is without prejudice to its rights and remedies following the outcome.

- If the instruction is for a change which makes it necessary to modify or alter the design of the Works, the contractor's consent is required (clause 3.9.1). This provision is often overlooked by employer and contractor alike. Its effect is twofold: it helps to impress on the employer that changes should be the exception rather than the rule, and it affords the contractor, as designer, the opportunity to resist a change which will result in serious amendment to its design. However, the contractor must not unreasonably delay or withhold its consent, for example to get even with an employer with whom it has had a difference on some other matter.
- The contractor need not comply with an instruction requiring a change under clause 5.1.2 (that is to say in respect of obligations as to access or use of parts of the site, limitations of working hours or space, or the carrying out and completion of the work in any specific order), to the extent that it makes a reasonable written objection to the employer. There are two points to note. Provided a reasonable objection is lodged, the contractor need not comply and it is not for the employer to decide what is reasonable. If the employer disputes the objection, it is a matter for the dispute resolution procedure. Therefore, such an objection can have expensive results – in delay to the Works for the employer who seeks dispute resolution under the contract and loses, and for the contractor in the costs of proceedings and delay if the employer wins. If the contractor objects, sweet reason dictates that the parties sit down and sort out the problem. The phrase 'to the extent' in clause 3.5 means that if the contractor objects to a part of the instruction, it is entitled to withhold compliance only from that part; it must comply with the rest of the instruction.
- If the contractor is the CDM co-ordinator, as is usual, it must notify the employer within a reasonable time of receiving the instruction, which requires a change or expenditure of a provisional sum. The contractor must state whether there is any objection to the instruction under the CDM Regulations. If so, it is for the employer to amend the instruction so as to remove the objection to the contractor's reasonable satisfaction. It is important to note that until the employer amends the instruction, clause 3.9.4 expressly provides that the contractor need not comply.

If the contractor, without proper grounds under the contract, does not carry out the instruction, clause 3.6 gives the employer the right to employ other persons to do whatever is necessary to carry out the instructions. Before so doing, the employer must give the contractor written notice requiring compliance with the instruction within 7 days. The employer's rights become operative if the contractor fails to comply. All costs incurred as a result may be deducted from the contract sum. This will include the money paid to the other persons and all other costs such as professional fees, if appropriate, and the cost of such things as scaffolding, cutting out and reinstatement. The wise employer will obtain alternative quotations for carrying out the work, unless time precludes it, so that it can be proved if necessary that the lowest practicable price has been paid: *Fairclough Building Ltd* v. *Rhuddlan Borough Council* (1985). The contractor is entitled to see a breakdown of the price. The employer may only deduct the additional cost of the contractor's failure to comply, excluding the amount the contractor would have had to be paid for doing the work anyway.

5.3.2 Specific instructions

The individual instructions empowered by the contract are discussed as follows.

Clause 2.10.1 In regard to discrepancy between Employer's Requirements and definition of the site boundary

The employer has a duty under clause 2.9 to define the site boundaries. In practice this will probably be done by the employer's solicitor. In this respect it should be noted that boundaries are notoriously difficult to settle with accuracy, particularly where old property is concerned. Once the employer has defined the boundary, presumably on a site plan, the contractor is entitled to proceed on the basis of that definition and if it commits an act of trespass solely as a result of a mistake in the employer's definition, the contractor would be able to look to the employer for an indemnity. If the parties agree that the contractor will provide the site, clause 2.9 must be amended so that it is the responsibility of the contractor to define the boundary. The contract empowers the employer to issue an instruction to correct a discrepancy between the Employer's Requirements and the definition of site boundary, which in turn entitles the contractor to a valuation. In the event that the contractor provides the site, this clause must also be amended.

Clause 3.12 To open up for inspection or carry out testing

This clause gives the employer power to instruct that work covered up is to be opened up for inspection, or to arrange for the testing of any materials whether or not they are already built into the Works. It is a valuable power, because its very existence can dissuade the contractor from attempting to incorporate work or materials which are not in accordance with the contract, and also because it enables the employer to check that work covered up is correct and the materials used are not cheap substitutes. The power is not without its drawbacks of course, and rightly so. If the workmanship or materials which are the subject of the check are found to be defective, the contractor is to stand the cost of putting matters right. If, however, everything is found to be in accordance with the contract, the employer must pay the cost of testing or opening up and making good. In addition, the contractor may have grounds for an extension of time and reimbursement of direct loss and/or expense under clauses 2.26.2.2 and 4.20.2.2 respectively.

Clause 3.13 Defective work, materials or goods

The employer has wide powers to issue instructions in respect of defects. Defective work or materials is work or materials which are not in accordance with the contract. The clause closely follows the equivalent clause in SBC. In the first edition of this book it was remarked that the employer, in addition to instructing the removal from site, had the power to instruct its rectification. This was a simple and obvious improvement on the position under JCT 98. There are clearly many instances when rectification rather than removal from site is indicated, for example, where there is defective paintwork. For some inexplicable reason, the JCT deleted that power and employers faced with, say, defective painting will be obliged to instruct removal from site instead of the more sensible rectification. It was thought that the reference to 'rectification pursuant to clause 8.4' in the then clause 12.1.1 was an oversight. However, the same reference to 'rectification' is retained in the new clause 5.1.1, therefore it is plainly incorrect. As an alternative or in addition to the power under clause 3.13.1, the employer, by clause 3.13.2, may issue an instruction requiring a change.

If the change is reasonably necessary as a result of the instruction to remove, the contractor is not entitled to any additional payment or extension of time, nor loss and/or expense. If the change is not entirely necessary, but only partly so, and the employer is taking the opportunity to change or incorporate other requirements, the contractor is entitled to payment, extension of time and loss and/or expense in respect of the part of the change instruction which is not reasonably necessary. There is a stipulation that the instruction may be issued only after consultation with the contractor. The intention is probably to allow the parties to agree on the extent of the instruction which is necessary. In practice, although the employer must consult, i.e. seek advice or opinion, there is no requirement that the employer must take heed of such advice or opinion.

A very valuable, but sometimes controversial, power is given to the employer by clause 3.13.3 which entitles the issue of instructions, under clause 3.12, to open up for inspection or to test other parts of the building or other materials. The idea is that the employer is entitled to establish to the employer's reasonable satisfaction whether there are any similar cases where work or materials are not in accordance with the contract. Whether or not the opening up or testing shows that similar defects exist, the contractor is not entitled to payment for carrying out the instruction and reinstatement. This is notwithstanding the provisions of the opening up clause 3.12 and the loss and/or expense clause 4.19. The contractor is entitled to an extension of time unless the work or materials were found not to be in accordance with the contract. Most importantly, before issuing the instruction, the employer must have had regard to a set of criteria in schedule 4 dubbed the 'Code of Practice'.

The intention of the Code is to assist in the fair and reasonable operation of clause 3.13.3. It provides that the employer and the contractor should try to agree the extent of the opening up or testing and the way it is to be accomplished, and the employer is to consider 15 criteria. They cover the kind of factors which might well give the employer cause for concern, for example, the importance of demonstrating that the failure is a one-off occurrence, the degree of significance of the failure in the context of the building as a whole and the implications on safety of a similar failure elsewhere, the standard of the contractor's supervision and any proposals which the contractor may make. In addition, the employer is to consider, as item 15, 'any other relevant matters', a category which could hardly be broader in this context. The instruction must be reasonable in all the circumstances. It appears, by use of the words 'to the extent', that if the instruction is not reasonable in the amount of opening up or testing required, the contractor is entitled to payment, extension of time and loss and/or expense for the part of the instruction which is not reasonable. What is or is not reasonable is a matter for the adjudicator or the arbitrator. It is likely that the employer has a great deal of scope in issuing instructions under this clause. The proviso that regard must be had to the Code of Practice would be satisfied if the employer simply read it before issuing the instruction. The effect of this provision, like its fellow in SBC, is very onerous so far as the contractor is concerned.

Clause 3.14 Workmanlike manner

Clause 2.1.1 requires among other things that all work must be carried out in a proper and workmanlike manner and in accordance with the construction phase plan. If the contractor fails to comply with that part of the clause, the employer, under clause 3.14, may issue an instruction, including requiring a change, if it is reasonably necessary as a consequence of the failure. This is very similar to the provisions of clause 3.13.2 following the discovery of work or materials not in accordance with the contract. In similar fashion, the contractor is not entitled to extension of time, loss and/or expense or any other payment. The only safeguard for the contractor is that the employer must consult it before issuing an instruction under this clause. That proviso is likely to be small comfort. When this clause was introduced, the Joint Contracts Tribunal indicated that it had taken account of the decision in *Greater Nottingham Co-operative Society Ltd* v. *Cementation Piling and Foundations Ltd and Others* (1988), which considered whether a nominated sub-contractor had liability to the employer in respect of bad workmanship. This clause should avoid the problem encountered in that case by making clear that, if the employer has to issue an instruction requiring a change on account of the contractor's failure to proceed in a workmanlike manner, the contractor can secure no advantage, financial or otherwise, as a result of such instruction.

Clauses 3.9.1 and 3.11 Effecting a change in the Employer's Requirements and expenditure of provisional sums

The employer is entitled to issue instructions to change its requirements under clause 3.9.1, but it is doubtful whether the employer is entitled to issue an instruction directly to vary the work or design. The consequence may be much the same, but not inevitably so. Clause 3.9.1 appears to leave the door ajar, but reference to 5.1, the definition of change, makes clear that only the requirements and not the design may be changed. For example, in the case of a hotel lobby, there may be lounge-type seating for 20 persons requested in the Employer's Requirements and provided in the Contractor's Proposals. The employer may wish to increase the seating capacity to 30 people. Under a traditional contract, the employer would inform the architect who would redesign that portion of the building so as to accommodate the increased seating requirement. The revised drawing showing exactly what was to be done would be issued to the contractor, together with an architect's instruction to carry out the revised work. Under DB, however, the employer could simply issue a change instruction to the contractor, stating that an additional ten seating spaces were required. In that instance, it would be for the contractor to look at the implications for the design and, if appropriate, refuse to comply (see also section 5.3.1). The contractor would only refuse, of course, if it was reasonable to do so.

It is difficult to state precisely what would be reasonable and each situation would be judged on its merits. In the example, the addition of a few seats seems hardly likely to provoke such a response from the contractor. It may be, however, that in order to accommodate the extra seats, the lobby would require enlarging, which might in turn create difficulties elsewhere. The contractor would probably be justified in withholding consent until such time as it could explain to the employer the full cost and other implications of the instruction, and possibly beyond that if the change involved a virtual redesign or massive rebuilding. If there was no problem, the contractor would simply take the instruction, carry out the redesign and, in accordance with schedule 1, submit it for comment and then proceed. The employer can exercise greater or less control over the result of the instruction by varying the degree of detail included in the instruction requiring a change. The employer might instruct 'ten extra seats in the lobby' or ' ten extra seats in the lobby, four of which should be facing the reception area and situated along the northern wall'.

Clause 3.9.4 applies if the contractor is also the CDM co-ordinator. It gives the contractor the right to object to an instruction in accordance with its duties under regulation 20 of the CDM Regulations. The contractor may only object under this clause if the instruction requires a change or if it concerns the expenditure of a provisional sum in the Employer's Requirements. The objection must be in writing. Once the objection is lodged, the employer must vary the terms of the instruction to remove the objection to the reasonable satisfaction of the contractor. Despite the provisions of clause 2.1, the contractor is not obliged to comply with the instruction until it has been so varied.

Clause 3.11 provides that the employer may issue instructions regarding the expenditure of provisional sums included in the Employer's Requirements. Note that the employer may not issue instructions in respect of any sums in the Contractor's Proposals, and if the contractor has included such sums, they must be transferred to the Employer's Requirements before the contract is executed. Although the contract provides for provisional sums, it is in the employer's best interests to include as few such sums as possible. Every sum introduces an element of uncertainty in price and time, which moves the risk towards the employer and away from the contractor roughly in proportion to the value of the provisional sum in relation to the contract sum.

Clause 2.35 Schedule of defects and requiring defects to be made good

Under the provisions of this clause, the employer is entitled to serve on the contractor, at the end of the rectification period, a list of defects which have appeared during that period, and also to issue such instructions as the employer considers necessary for the correction of defects during the period. These clauses are dealt with in detail in Chapter 7.

Clause 3.10 Postponing work

This clause provides a valuable power to the employer to postpone not only the carrying out of the Works, but also the design of the work or any part. This reflects the contractor's responsibilities under this form of contract. This power must be exercised with caution. Postponement entitles the contractor to an extension of time under clause 2.26.2.1, direct loss and/or expense under 4.20.2.1 and to terminate its employment under clause 8.9.2 if the postponement affects substantially the whole of the Works for a period exceeding the period entered in the contract particulars. It is worth noting that a court or arbitrator may decide that a postponement instruction has been issued even though the employer has not used those words or referred to this clause. A court will look at whether a letter or instruction issued by the employer amounts to a postponement instruction, although issued for some other purpose: *Holland Hannen & Cubitts (Northern) Ltd* v. *Welsh Health Technical Services Organisation and Another* (1981) and *M Harrison & Co (Leeds) Ltd* v. *Leeds City Council* (1980). Although not without doubt, it seems unlikely that a contractor can look to benefit from a postponement instruction which arises as a result of some defect in the work: *Gloucestershire County Council* v. *Richardson* (1967).

Clause 3.16 Regarding antiquities, including excavation, examination or removal by third parties

Under this clause, the employer has power to instruct the contractor to take specific action in respect of antiquities which it has reported under the provisions of clause 3.15.3. Notwithstanding the general nature of the employer's power under this clause, it is specified that the instructions may require the contractor to permit the examination, excavation or removal of the object by a third party. Although the third party is deemed to be an employer's person for the purposes of clauses 6.1 and 6.2 in respect of injury to persons and property and indemnity to the employer, it is not envisaged that the third party will be employed by the employer under clause 2.6 and, therefore, provision for loss and expense and extension of time is made in clauses 3.17 and 2.26.2.1, respectively.

5.4 Powers

5.4.1 General

Under traditional forms of contract, the employer has few express rights of importance other than the obvious right to receive the building, completed in accordance with the contract, on the due date. Under this form, however, in addition to the employer's powers to issue instructions, the contract confers some other substantial powers which are worthy of mention. A power exists whenever the contract states that the employer may do something. The employer's powers are summarised in Fig. 5.1.

5.4.2 Access

Clause 3.1 is one of only two clauses which expressly refer to the employer's agent. It provides that the employer's agent and any person authorised by the employer or the agent must have access to the Works, workshops or any other places where work is being prepared for the contract. The contractor is obliged to insert a term in appropriate sub-contracts so as to obtain similar rights for the employer and the employer's representatives. The contractor must go further and do everything reasonably necessary in order to give effect to those rights. The employer may take advantage of the powers under this clause at all reasonable times. In this context a reasonable time would be during normal working hours. There is just one important proviso: the contractor and the sub-contractor may impose whatever reasonable restrictions are necessary in order to protect their proprietary rights in the work to which the employer has access. This is a vital safeguard at a time when increasing amounts of building components are of a specialist nature and trade secrets must be safeguarded.

If the employer is to appoint a clerk of works in addition to the employer's agent, the clerk of works should be authorised to act under this clause as well as being listed, with duties, in the Employer's Requirements.

5.4.3 Partial possession

The employer's power to take part of the Works into possession before practical completion of the whole is governed by clauses 2.30–2.34. Although many employers seem to assume that this power is unfettered and subject only to their wishes, such is not the case. The contractor has power to refuse consent. Consent must not be unreasonably withheld, but in practice it can be very difficult for the employer to maintain that the refusal is unreasonable in the face of the contractor's insistence that partial possession will hamper its progress. Of course, generally the contractor will be delighted to secure the advantages which flow from partial possession (see Chapter 7). This clause is not intended for use where the employer knows before tenders are invited that possession of the building in parts will be required. Completion of the Works in sections, with the appropriate parts properly identified in the contract particulars, should be employed for that purpose. The partial possession provision is intended for use only where the employer decides during the progress of the Works that partial possession of a particular part is desired.

5.4.4 Effect insurance

The insurance provisions are noted in detail in Chapter 11. It should be noted that the employer has important powers to scrutinise insurance policies and to take out and maintain insurance if the contractor fails to do so. Under schedule 3, paragraph C.4.4, the employer may terminate the contractor's employment if it is just and equitable following the discovery of loss or damage to the work or site materials caused by any of the insured risks. For a fuller discussion of this provision see Chapter 12.

5.4.5 Deferment of possession

If the employer fails to give possession on the date specified in the contract, it is normally a serious breach of contract: *Freeman & Son* v. *Hensler* (1900). Normally, the consequences are damages for the breach and possibly repudiation if the failure is severe. Although it is so serious, there are numerous instances where the employer offends in this way. Sometimes, employer and contractor agree informally that possession may be late, but strictly a special agreement should be entered into. Clause 2.4, therefore, gives the employer an important power to defer the giving of possession for a period which must not exceed 6 weeks, but may be whatever lesser period is stated in the contract particulars. The employer must state in the contract particulars whether this provision is to apply. In view of the consequences of failure to give possession on the due date, it is considered vital that the contract particulars should state that this clause should apply unless the employer is absolutely certain that there can be no difficulties. It is thought unlikely that contractors increase their tenders significantly, if at all, to take account of the risk.

Of course, deferment has its consequences too, but they are regulated by the contractual machinery and provided the deferment does not exceed the stated period, the contractor will be entitled to an extension of time under clause 2.26.3 and loss and/or expense under clause 4.20.5. Since the introduction of the catch-all clauses 2.26.5 and 4.20.5 for extension of time and loss and/or expense respectively, it may be thought that even a long deferment can be managed under the contract. Although it is correct to say that the contract mechanism is available, it should not be forgotten that the contractor has the option to claim damages for breach of contract, if the deferment exceeds 6 weeks, instead of accepting redress under the contract. If the deferment is lengthy, the contractor may be able to treat it as a repudiation of the employer's obligations.

5.4.6 Deduction of liquidated damages

If the contractor fails to complete the Works or any section by the date for completion stated in the contract particulars or by any extended date, the employer is entitled to recover liquidated damages at the rate stated in the contract particulars. This can have severe consequences for the contractor, especially where the amount of liquidated damages is fixed as a substantial sum per day or per week. The exercise of the power is circumscribed by three very important preconditions: the employer must have issued a non-completion notice under clause 2.28 that the contractor has failed to complete by the due date, the employer must have given notice of an intention to recover liquidated damages and the employer must have issued an appropriate withholding notice. If a new date for completion is fixed, the non-completion notice is automatically cancelled and the employer must issue a new notice. Any amount of liquidated damages which has been recovered must be repaid, but there is no requirement for interest. If, however, the employer has recovered liquidated damages unlawfully, for example because the liquidated damages clause is defective or the notice of non-completion has not been issued, the contractor may have a claim for recovery of interest as special damages for the breach: *Department of the Environment for Northern Ireland* v. *Farrans (Construction) Ltd* (1982).

In *A Bell and Son (Paddington) Ltd* v. *CBF Residential Care and Housing Association* (1989), the judge confirmed that both notices of non-completion and a written requirement for payment were preconditions. This clause has been substantially redrafted to comply with the Housing Grants, Construction and Regeneration Act 1996 Part II and a full discussion is to be found in section 8.4.2.

It is important to remember that the employer or the employer's agent, even if an architect, has no duty to act fairly between the parties in issuing the non-completion notice on which the recovery of liquidated damages is based. Such a notice is not of binding effect until arbitration, as would be a certificate of similar content by the architect under the provisions of SBC: *J F Finnegan Ltd* v. *Ford Seller Morris Developments Ltd (No.1)* (1991). Thus, if disputed, it seems that liquidated damages could not be recovered from the retention fund until arbitrated upon.

5.4.7 Review extensions of time

Under the provisions of clause 2.25.5, the employer is empowered to review the extension of time situation and either confirm the date for completion previously fixed or fix a new date earlier or later than the previous date. This valuable power can be used by the employer to prevent time becoming at large. For a further discussion see Chapter 8.

5.4.8 Work not forming part of the contract

The contractor has the right to exclusive possession of the site while it is carrying out the building contract. This is subject to some exceptions (see Chapter 7). If it were not for the inclusion of clause 2.6 in the contract, the employer would have no right to have work carried out by other persons on the site until the contractor had finished. Apart from legalities, it makes sense that the contractor would not be able to get on and complete its work properly and within the contract period if constantly interrupted by other persons tramping across the area where it is working. Employers, however, frequently wish to engage others to do particular parts of the work and they want far more control over these persons than would be the case if they were simply named sub-contractors under schedule 2, paragraph 2 (see Chapter 6).

Clause 2.6 gives the employer power to engage others for particular work subject to certain conditions. There are two situations envisaged by the contract:

- If the employer has included in the Requirements sufficient information about the particular work so as to enable the contractor to complete the contract Works in accordance with the contract, the contractor must allow the work to be carried out by others (clause 2.6.1).
- If the employer has made no reference to the particular work in the Requirements, but wishes to have such particular work carried out by others, the employer may arrange for the carrying out of the work if the contractor consents. There is a stipulation that the contractor must not unreasonably withhold its consent (clause 2.6.2).

In a contract whose philosophy is to place as much responsibility as possible in the hands of the contractor, a clause like this seems rather out of place. Clearly, the contractor will be able to organise resources, hit targets and generally manage the project more effectively if it has complete control over the site and resources. Employers, therefore, would be advised to consider very carefully whether they really want to exercise the power contained in this clause which will almost certainly affect some element of the project, whether financial or concerning time or quality. If the employer feels it necessary to use this clause, there will usually be a price to pay. Clauses 2.26.5 and 4.20.5 entitle the contractor to an extension of time and loss and/or expense respectively in appropriate cases (see Chapter 8).

Clause 2.6 refers to 'work not forming part of the contract'. That is precisely correct. The work referred to in this clause is not work which the employer can require the contractor to carry out, even, it is thought, by using the powers in clause 3.9. Neither can this clause be used to allow the employer to omit work from the contract under clause 3.9 and give it to others using clause 2.6. This is particularly so when the person concerned is not an independent professional but the person directly interested: *Vonlynn Holdings Ltd* v. *Patrick Flaherty Contracts Ltd* (1988); *AMEC Building Ltd* v. *Cadmus Investment Co Ltd* (1997).

5.5 Employer's duties

5.5.1 General

Duties are normally indicated in the contract clauses by the use of the word 'shall'. If the employer fails to carry out any duties under the contract, it will be a breach of contract for which the contractor will have a remedy in damages (which may, of course, be nominal) quite apart from any specific remedy prescribed under the terms of the contract itself. Some of the employer's most important duties are discussed below. A full list of those duties is given in Fig. 5.2.

5.5.2 Notices

Since there is no independent architect administering the contract, it falls to the employer or the employer's agent to issue any notices required under the contract. The most important notices relate to discrepancies which the employer may discover under clauses 2.10.2 and 2.15.1, practical completion statement under clause 2.27, notices of completion or making good under clause 2.36 and non-completion notice under clause 2.28. A full list of the notices and statements to be issued by the employer is given in Fig. 5.4. The effect of such notices is discussed in detail in the appropriate chapter, but it should be appreciated that if the employer neglects to issue a notice at the right time, the consequences will always be serious.

5.5.3 Possession

The employer must give possession of the site to the contractor on the due date unless the deferment clause 2.4 applies and it has been properly operated by the employer.

5.5.4 Extensions of time

The employer has a duty to make extensions of time in accordance with the procedures set out in clauses 2.23–2.26 (see Chapter 8). This task would normally be undertaken by the architect if a traditional contract was used. Since the employer or the employer's agent can in no sense be considered to be disinterested, it is suggested that the fixing of a new date for completion under this form of contract will be examined by the courts or an arbitrator with correspondingly more care than would be the case under a traditional form.

Clause	Statement or notice
2.5.1	Notice to insurers if schedule 3, paragraphs B or C.2 apply and employer wishes to occupy or use the site or the Works.
2.10.2	Notice of divergence between Employer's Requirements and definition of site boundary.
2.14.1	Notice of decision regarding treatment of discrepancy within the Employer's Requirements.
2.15.1	Notice of divergence between statutory requirements and the Employer's Requirements or the Contractor's Proposals.
2.25.2	Notice of employer's decision regarding extension of time.
2.25.4	Notice fixing a completion date earlier than previously fixed.
2.25.5	Notice after the completion date, fixing a later date or fixing an earlier date or confirming the date previously fixed.
2.27.1	Practical completion statement.
2.27.2	Section completion statement.
2.28	Non-completion notice. Further non-completion notice.
2.29.1.2	Notice that liquidated damages may be required or deducted.
2.29.2	Notice that liquidated damages are required or are to be deducted.
3.6	Notice requiring compliance with an instruction.
3.19	Notice of replacement of CDM co-ordinator or principal contractor.
4.10.3	Notice specifying the amount to be paid.
4.10.4	Notice of intention to withhold payment.
4.12.5	Notice if contractor does not submit the final account and final statement.
4.12.8	Notice specifying the amount to be paid in respect of the final account.
4.12.9	Notice of intention to withhold payment of amount from final account balance.
6.1.2	Notice specifying divergence between statutory requirements and Employer's Requirements or Contractor's Proposals.
6.10.2	Notice either that despite cessation of terrorism cover the Works are to continue or that the contractor's employment is to terminate.
7A.1	Notice regarding Purchaser's and Tenant's (P&T) rights.
7B.1	Notice regarding funder's rights.

Fig. 5.4 Statements and notices to be given by the employer.

7C	Notice regarding contractor's warranty to P&T.
7D	Notice regarding contractor's warranty to funder.
7E	Notice regarding sub-contractor's warranty to P&T, funder or employer.
8.4.2	Notice of termination.
8.5.1	Notice of termination after insolvency.
8.6	Notice of termination due to corruption.
8.8.1	Notice that Works are not to be completed.
8.11.1	Notice giving 7 days to end suspension. Further notice of termination.
9.3	Joint notice regarding amended CIMAR.
9.4.1	Notice of arbitration.
Schedule 2, para. 5.4	Notice accepting estimate; *or* wishing to negotiate or in default referring to dispute resolution procedures; *or* applying clause 4.19.
Schedule 3, para. C.4.4	Notice of termination.

Fig. 5.4 *Continued*

5.5.5 Payment

Perhaps the most important duty of the employer is to pay the contractor in accordance with the terms of the contract (see Chapter 10). Failure to pay is dealt with in clause 8.9.1.1 where the contractor has power to terminate its employment under certain conditions. The contractor has no right at common law to stop work just because it has not been paid what it considers to be the correct amount: *Lubenham Fidelities & Investment Co* v. *South Pembrokeshire District Council and Wigley Fox Partnership* (1986). This can be a very serious matter for the contractor who may not be able to fund continuation of the project in the face of the employer's breach. In *Lubbenham,* the contractor was unable to terminate under the contract provisions because the employer had correctly paid the amount shown on the architect's certificate. It was the certified amount which was clearly and demonstrably wrong. There are no certificates under DB and the contractor is potentially in a strong position. It should also be noted that the courts do appreciate the contractor's problems where payment is withheld, and such withholding may be held to amount to a repudiatory breach if it is so repeated that the contractor has no realistic expectation that it will ever be paid (*D R Bradley (Cable Jointing) Ltd* v. *Jefco Mechanical Services* (1989)) or even perhaps if the employer simply threatens to make no further payment until the project is finished: *C J Elvin Building Services Ltd* v. *Peter and Alexa Noble* (2003). However, under the provisions of section 112 of the Housing Grants, Construction and Regeneration Act 1996, the contractor has the right to suspend performance in such circumstances on 7 days' notice (clause 4.11).

Payment by cheque is probably good payment, although in theory the payment is not made until the cheque is cleared through the bank. Contractors faced with an employer who simply does not pay are in serious difficulties. An employer can often resist payment by demonstrating a strong *bona fide* case for resisting payment: *C M Pillings & Co Ltd* v. *Kent Investments Ltd (1985); R M Douglas Construction* v. *Bass Leisure Ltd* (1991). However, the excellent and rapid process under the adjudication procedure in clause 9.2 could ensure speedy attention to the problem (see Chapter 13).

If the contractor opts not to seek adjudication or to suspend the work, the contractor's best way forward is to operate the termination provisions, provided it is sure that the failure can properly be brought under this clause. On receipt of the default notice, the employer may immediately pay. If the employer does not pay within the stipulated period, the contractor can terminate its employment and it may be able to negotiate suitable terms for continuance of work thereafter. If terms are not agreed, the employer must settle the account and if the employer fails to do so, referral to adjudication or arbitration as appropriate can proceed without the contractor having the burden of carrying out the work.

Chapter 6
Sub-contractors and Statutory Requirements

6.1 General

This chapter considers statutory requirements, assignment and sub-contracts, execution of work not forming part of the contract, third party rights and collateral warranties. There is no provision for nomination under this contract, but the supplemental provisions allow for 'Persons named as sub-contractors in Employer's Requirements'. How far such provisions permit the employer to impose a choice of sub-contractor upon the main contractor will be considered later in this chapter.

Sub-contracting, assignment and novation are often confused. Before considering the contract provisions in detail, it is important to understand the difference between these terms. Conveniently, they were set out with admirable clarity by Lord Justice Staughton in *St Martins Property Corporation Ltd and St Martins Property Investments Ltd* v. *Sir Robert McAlpine & Sons Ltd and Linden Gardens Trust Ltd* v. *Lenesta Sludge Disposals Ltd, McLaughlin & Harvey plc and Ashwell Construction Company Ltd* (1992) in the Court of Appeal:

> '(a) Novation This is the process by which a contract between A and B is transformed into a contract between A and C. It can only be achieved by agreement between all three of them, A, B and C. Unless there is such an agreement, and therefore a novation, neither A nor B can rid himself of any obligation which he owes to the other under the contract. This is commonly expressed in the proposition that the burden of the contract cannot be assigned, unilaterally. If A is entitled to look to B for payment under the contract, he cannot be compelled to look to C instead, unless there is a novation. Otherwise B remains liable, even if he has assigned his rights under the contract to C . . .
>
> (b) Assignment This consists in the transfer from B to C of the benefit of one or more obligations that A owes to B. These may be obligations to pay money, or to perform other contractual promises, or to pay damages for a breach of contract, subject of course to the common law prohibition on the assignment of a bare course of action. But the nature and content of the obligation, as I have said, may not be changed by an assignment. It is this concept which lies, in my view, behind the doctrine that personal contracts are not assignable . . . Thus if A agrees to serve B as chauffeur, gardener or valet, his obligation cannot by an assignment make him liable to serve C, who may have different tastes in cars, or plants, or the care of his clothes . . .
>
> (c) Sub-contracting I turn now to the topic of sub-contracting, or what has been called in this and other cases vicarious performance. In many types of contract it is immaterial whether a party performs his obligations personally, or by somebody else. Thus a contract to sell soya beans, by shipping them from a United States port and tendering the bill of lading to the buyer, can be and frequently is performed by the seller tendering a bill of lading for soya beans that somebody else has shipped.'

6.2 Sub-contractors

6.2.1 Assignment

Clause 7.1 contains the usual restriction on the assignment of the contract by either party without the written consent of the other. In the *St Martins* case, the House of Lords (1993) held that this clause effectively prevents the benefit of the contract being assigned. For example, the employer might wish to sell the building before the final account and final statement become conclusive, or the contractor may wish to assign the right to receive payment in return for a cash advance. With clause 7.1 in place, consent would have to be given by the other party in each case, but see clause 7.2 considered below. There is nothing to prevent a party refusing consent on grounds which might be considered unreasonable. This can pose real problems and if the employer might possibly wish to assign the benefit of the contract (i.e. sell or otherwise transfer the property to another) before practical completion, an amendment to the clause is indicated.

The general law forbids assignment of the burden of a contract, so this clause is superfluous in that regard. For the contractor to effectively transfer to another the duty to carry out the work set out in the contract documents, or for the employer to transfer the duty to pay for such work, would require the consent of both parties to the contract, together with the consent of the party who is to shoulder the burden in place of either contractor or employer.

Clause 7.2 will apply unless otherwise stated in the contract particulars. It contemplates the situation where the employer sells the freehold or the leasehold interest in the premises comprising the Works to a third party, or where the employer grants a leasehold interest in the premises. In any of these instances, the employer may assign to that third party the right to bring proceedings in the employer's name and to enforce any of the terms of the contract. There is a proviso that the third party cannot dispute any agreement which is legally enforceable and which is entered into between the employer and the contractor before the assignment. This clause does not give the employer the right to sell the premises before receiving them from the contractor at practical completion, therefore it does not conflict with clause 7.1. However, once the employer has received the building and disposed of it by sale or lease, it enables the purchaser to act as if he or she was the employer so far as the benefits of the contract are concerned. For example, the obligation to pay the contractor remains with the employer, but the purchaser can enforce the defects provisions. However, if the employer and the contractor have entered into an agreement under which the contractor is not obliged to make good certain defects and no monetary deduction is to be made, it is binding on the purchaser of the premises under clause 7.2.

6.2.2 Sub-letting

An important difference between this contract and SBC is that clause 3.3.2 allows the contractor to sub-let design provided it has the written consent of the employer. In contrast to assignment, the employer may not unreasonably delay or withhold consent. If the employer does give consent, the contractor's obligations under clause 2.17 are not affected (see section 3.2). In practice, the employer will normally give consent readily, because few contractors keep high calibre design teams on the staff, and an employer may have selected a particular contractor partly on account of the prestigious design team it has assembled. If that is the situation, the contractor would be well advised to have a suitable amendment made to the printed form to allow such sub-letting without the requirement for consent.

Professional fee agreements can be ill-suited to the design and build concept. Even such things as reference to the 'client' have to be understood in a new light as meaning the contractor. Therefore, references to the 'contractor' become confusing. Many contractors who regularly carry out design and build contracts have their own standard forms of agreement for the design team. Those who do not have such forms should seriously consider acquiring them. The draft terms of engagement produced by the RIBA were briefly considered in section 3.3.

Clause 3.3.1 is a prohibition on the sub-letting of the whole or any part of the Works without the written permission of the employer. Once again, such consent must not be unreasonably delayed or withheld. There is no requirement that the contractor must inform the employer of the names of sub-contractors. It is merely consent to the fact of sub-contracting which is required. It would be reasonable for the employer to refuse to give consent until the name and perhaps other details of the prospective sub-contractor were made known. To avoid dispute, the employer could state in the Requirements that such information has to be submitted before consent will be given. The last part of the clause states that the contractor remains wholly responsible for the carrying out and completion of the Works in accordance with clause 2.1, even though some or even the whole of the Works is sub-let. This clause was inserted in most JCT contracts in an excess of caution following the decision in *Scott Lithgow* v. *Secretary of State for Defence* (1989).

The employer has no contractual relationship with the sub-contractor, and must look to the contractor if there is any defect in the sub-contractor's work. This is so even if the sub-contractor has gone into liquidation. The contractor will still remain responsible for its work. It is known as the 'contractual chain'. At times such a chain can be a long one such as when sub-sub- or even sub-sub-sub-contractors are involved. Responsibility for defects goes right up through the chain and redress goes down through the chain. If the chain is broken by insolvency, responsibility rests with the party on the side of the insolvency nearest the employer (see Fig. 6.1). If the contractor has become insolvent when a defect is found in a sub-contractor's work, the employer has no contractual remedy and, following *Murphy* v. *Brentwood* (1990), little hope of a tortious remedy. To overcome such problems, the employer may make the provision of an acceptable warranty on the part of the sub-contractor a precondition to the giving of any consent to sub-letting. Alternatively, the employer may insert such a stipulation, accompanied by an example of the warranty required, in the Requirements. This and other warranties are considered in section 6.6 below.

Clause 3.3.3 provides that clauses 3.3 and 3.4 do not apply to work by statutory undertakers. This ought to be obvious – on the basis that statutory undertakers are not usually acting as sub-contractors. Of course, where a statutory undertaker is acting as a sub-contractor, clauses 3.3 and 3.4 will apply. Clause 3.4 sets out certain provisions as a condition to sub-letting. Clause 3.4.1 states that each sub-contract must provide that the employment of the sub-contractor terminates immediately termination of the contractor's employment takes place. Clause 3.4.2

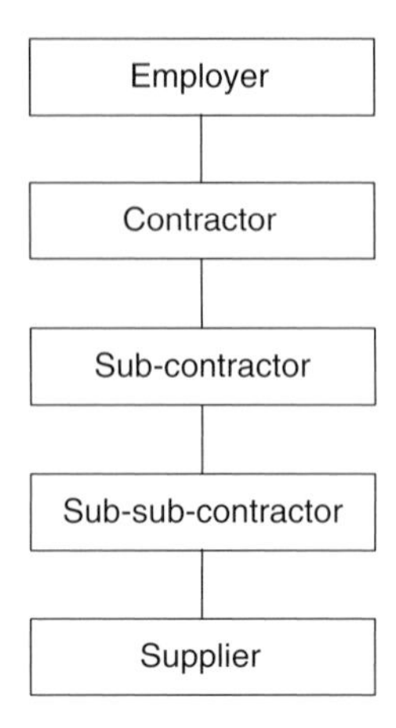

Fig. 6.1 Contractual chain – if sub-sub-contractor becomes insolvent, liability rests with the sub-contractor.

states that the sub-contract must include certain provisions regarding unfixed materials and goods delivered to the Works or adjacent to the Works. There are three terms:

- Such materials and goods must not be removed without the contractor's consent unless for use on the Works (clause 3.4.3 makes the contractor's consent subject to the employer's consent).
- Ownership of such materials and goods is to be automatically transferred to the employer after the value has been included in an interim payment.
- If the contractor pays for such materials and goods before itself being paid by the employer, ownership passes to the contractor.

The operation of these clauses is not to affect ownership in off-site materials passing to the contractor, as provided in clause 4.15.2.1 of DB. Other clauses which must be included deal with rights of access to sub-contractors' workshops, the sub-contractor's right to interest on late payments and sub-contractor warranties.

The employer should ensure that the relevant provisions are included in sub-contracts and perhaps refuse to consent to sub-contracting unless evidence of such inclusion is produced. Although standard sub-contract DBSub/C contains such provisions, many contractors habitually sub-contract using their own terms, which not only do not contain such provisions but also do not create a satisfactory 'back to back' sub-contractual arrangement. Even if such provisions are included, they are ineffective to safeguard the employer from the perils of retention of title, if the sub-contractor has bought the goods itself on terms that the supplier retains ownership until payment is made. Building contract chains are so long that it is virtually impossible to check down to the ultimate supplier that ownership has passed unimpeded up to the contractor.

Breach of the provisions of clauses 3.3 or 7.1 is sufficient grounds for termination by the employer under clause 8.4.1.4 (see Chapter 12).

6.2.3 Persons named as sub-contractors in Employer's Requirements (schedule 2, paragraph 2)

This provision gives the employer some assurance that certain parts of the project can be carried out by a sub-contractor of the employer's own choice. It has similarities to the naming provisions in IC, ICD and ACA 3. The provision is only to apply if the Employer's Requirements state that certain work is to be carried out by a named person employed as a sub-contractor by the contractor. The work is termed 'named sub-contract work' and the sub-contractor is termed a 'Named Sub-Contractor'.

Paragraph 2.1.1 stipulates that the contractor must enter into a sub-contract with the named person as soon as reasonably practicable after entering into the main contract. That is to say that the contractor must enter into the sub-contract as soon as it can in practice, or allowing for the current situation. As soon as it has entered into the contract, the contractor will have a hundred and one things to do immediately – things which cannot be done before the main contract is executed. This sub-contract will be only one of those things. It is always in the contractor's interest to execute sub-contracts promptly so as to safeguard the price. Contractors often have a very loose arrangement with prospective sub-contractors even though the tender price may be based on figures provided by them. For example, there is nothing in law to prevent a sub-contractor from withdrawing its price before acceptance, even if it has stated in its quotation that it will remain open for a specific period (unless the contractor has paid to keep the option open). If this happened in the case of a named person, the result could be disastrous. At best, the contractor could be forced into the position of taking a loss on the sub-contract. The contractor must notify the employer of the date of the sub-contract. There is no requirement that this notification should be in writing, but it is sensible for the contractor to serve all notices in writing.

The contractor may not be able to enter into the sub-contract, perhaps because the named person has withdrawn its quotation and refuses to submit another price. Whatever the reason, the contractor must notify the employer immediately to allow as much time as possible for action to mitigate the effect of the problem. The contractor must give the reason for its inability to conclude a sub-contract. If the reason is *bona fide*, the employer may take alternative courses of action:

- If the reason is connected to the item in the Employer's Requirements, the employer may issue a change instruction to remove the reason; or
- The employer may issue a change instruction to omit the named sub-contract work from the Requirements. If the employer chooses the second option:
- The employer may issue a further instruction requiring the contractor to select another person to carry out the work subject only to the employer's reasonable approval of that person. There seems no reason in principle why the contractor should not choose itself in an appropriate case.
- The employer may state that the work is to be carried out by a directly employed person as referred to in clause 2.6 (see section 6.5).

The employer may not name a substitute person in the change instruction.

Whether or not the reason advanced by the contractor for inability to enter into the sub-contract is *bona fide* need not be left to the opinion of the employer nor to the employer's agent. If a dispute develops over the point, it may be referred to adjudication under the provisions of clause 9.2.

The sub-contract termination situation is dealt with in paragraphs 2.1.5–2.1.8. The contractor may not terminate the named sub-contractor's employment for default unless it first obtains the employer's consent, which must not be unreasonably withheld or delayed (paragraph 2.1.5). If the contractor proceeds to carry out the termination, it is responsible for completing whatever work is left unfinished. Although the employer is not to issue a change instruction to cover the situation, the work required to complete is to be treated as though it resulted from a change instruction. There are two exceptions:

- If the termination is the result of the contractor's default, i.e. if the sub-contractor terminates; or
- If the contractor has not obtained the employer's consent under paragraph 2.1.5.

Paragraph 2.1.7 is obscurely drafted, but it seems that the contractor must pay the employer any amounts which it has recovered, or which it could have recovered from the defaulting sub-contractor, using reasonable diligence. What qualifies as reasonable diligence will vary with the circumstances. The amounts to be recovered are those which are legally due to the contractor as a result of the termination to reduce the cost of completion. This will only apply if the termination is due to the sub-contractor's default. To avoid the sub-contractor successfully contending in any proceedings that the contractor has suffered no loss and, therefore, that it has nothing to recover, the contractor is obliged by paragraph 2.1.8 to insert an appropriate clause in the sub-contract. The clause must state that the sub-contractor, having notice of paragraph 2, undertakes not to contend that the contractor has suffered no loss and that its liability should be reduced or extinguished in any way. It is thought that such a clause would be effective in practice: *Haviland and Others* v. *Long and Another, Dunn Trust* (1952).

6.3 Statutory requirements

Statutory requirements are no longer dealt with under one clause. The provisions previously found in clause 6 of WCD 98 are now scattered among clauses 2.1.1, 2.1.2, 2.1.3, 2.15, 2.16 and 2.18. Although the provisions are modelled on the equivalent clause in SBC, there are important additions and changes which reflect the particular philosophy of this contract. The contractor's principal obligation is contained in clauses 2.1.1 and 2.1.3. Clause 2.1.1 will apply except to the extent that the Employer's Requirements specifically state that they comply with statutory requirements (clause 2.1.2). For example, if the Employer's Requirements say that the particular requirements as stated for an auditorium comply with statutory requirements, the contractor is entitled to assume that they do so comply and it may complete the design on that basis. If there is later found to be a failure to comply, the employer and not the contractor will be responsible for the cost of correcting the error. The contractor is not entitled to assume compliance for any part of the building other than the specific part stated in the Requirements – in this case, the auditorium. It should be noted that this sets out the position between the employer and the contractor, but the contractor cannot shelter behind this proviso so far as the statutory authorities are concerned. It would seem, however, that the provision would be sufficient, in most instances, to enable the contractor to recover from the employer any costs incurred in making the work covered by clause 2.1.2 comply with statutory requirements.

The contractor must comply with statutory requirements, which are defined in clause 1.1 to include any Act of Parliament, instrument, regulation, rule or order or any regulation or bylaw of a local authority or statutory undertaker who has jurisdiction with respect to the Works or with whose systems the Works will be connected (water, electricity and drainage systems are obvious examples). That is said to include development control requirements and the contract refers to all these regulatory matters as statutory requirements. Development control requirements are defined in clause 1.1 as 'any statutory provisions and any decision of a relevant authority thereunder which control the right to develop the site'. They are often referred to for ease as 'planning requirements' and in a narrow sense that is correct. However, there are other statutory provisions which control the right to develop the site in a particular way. Examples of such provisions are the Fire Regulations and the requirement for an entertainment licence for an appropriate development. Thus, for a particular development, the failure to satisfy any statutory provisions which determine whether a site can be developed for a particular purpose would amount to a failure to satisfy development control requirements.

It is clear that the contractor's obligation is to comply with statutory requirements, not only in the design it is to complete but also in the whole of the construction, including construction of part of the design which may already be in the Employer's Requirements. Under clause 2.1.3, the contractor must give the employer all statutory approvals received. It must also submit all notices in connection with statutory requirements (clause 2.1.1).

Other than where the Employer's Requirements specifically state otherwise, it is clear that the contractor bears responsibility for compliance. This is emphasised by clause 2.15 which explains the procedure if either the employer or the contractor finds a divergence between statutory requirements and either the Employer's Requirements (which includes any changes) or the Contractor's Proposals. Whoever finds the divergence must immediately notify the other in writing stating the divergence. Then the contractor must write to the employer with its proposals for dealing with the divergence at its own cost. The employer may not unreasonably delay or withhold its consent and the contractor must complete both the design and construction according to the amendment. Provided that the amendment deals with the divergence, it seems that the employer will have no grounds to withhold consent. An employer who dislikes the proposed amendment must issue a change instruction and pay the cost. The contractor is not required to pay the cost if the situation is covered by clause 2.15.2 (see below). There is a very curious provision: the employer must note the amendment on 'the documents referred to in clause 2.7.1'. The documents referred to are the contract documents. Leaving on one side the fact that it would have been easier to refer to the 'Contract Documents' rather than the present convoluted formula, it is clear that the provision allows the employer to amend the contract documents to correct a divergence after the contract has been executed. It seems likely that the intention is to prevent the contractor being able to claim payment later on the grounds that the work represented in the amendment is not included in the contract. It not only appears to be unnecessary, it can be dangerous. It is not clear why the employer should be able to amend the contract documents on account of a divergence caused by a change instruction. In the last edition of this book it was pointed out that this provision could bear some adjustment; it remains unaltered. Although under clause 5.1, a change may mean a change in the Employer's Requirements which necessitates an alteration in the design or quality or quantity of the Works, the alteration to the design in response to the instruction is entirely a matter for the contractor. It is entitled to be paid for correctly responding to the instruction. Therefore, if there is a divergence between the design response to a change instruction and statutory requirements, the responsibility must lie with the contractor.

Clause 2.15.2 is in three parts and it appears to modify the contractor's obligations under certain circumstances, although the drafting is not such as to encourage understanding by the people who will be using it.

Clause 2.15.2.1: If after the base date there is a change to statutory requirements (which include development control requirements) which makes it necessary to amend the Contractor's Proposals, the amendment is treated as a change. This means that the contractor will be entitled to payment for the change under clause 5.

Clause 2.15.2.2: If after the base date there is a decision made by the appropriate authority for development control requirements and the decision sets out terms of a permission or approval which make it necessary to amend the Contractor's Proposals, the amendment is treated as a change. This is subject to anything that the employer may have stated to the contrary in the Requirements. This last provision is intended to enable the employer to place the risk of satisfying development control requirements on the shoulders of the contractor. On its own, it is doubtful that it is enough to carry out that purpose.

Clause 2.15.2.3: If it becomes necessary to amend that part or the whole of the Employer's Requirements which the employer has stated specifically comply with statutory requirements, the employer must issue a change instruction for the purpose.

The intention behind the drafting of these clauses appears to be that, if after the base date there is a change in statutory requirements, the contractor is entitled to be paid the cost of dealing with the change. For example, there could be an amendment to the Health and Safety at Work Act or to the Building Regulations or to one of the Planning Acts, which affects the building for which the contractor has already entered into a contract. The contractor will be obliged to comply with the change in statutory requirements, but it is paid as though the employer had issued an instruction requiring that change. The same approach is applied to a permission or approval given after base date. The most common situation will be a decision given about reserved matters in a planning approval. The local authority may attach conditions and the contractor has to make amendments to its proposals to satisfy them. Once again, the amendment is treated as though the employer gave an express instruction for it and the contractor is entitled to be paid accordingly. There is provision for the employer to make the contractor take that risk by an appropriate statement in the Employer's Requirements. This is unlikely to be popular with contractors, because it is asking the contractor to budget for the unknown. There is a great difference between statutory requirements such as Building Regulations and development control requirements such as planning requirements. Whether or not a building will satisfy the Building Regulations is a matter which the contractor should be able to determine when it designs. The contractor may commence building after simply serving the appropriate notice on the local authority. It is not possible to say whether a building will be given planning permission either in whole or in respect of any part and until permission is obtained, the contractor may not commence building. The last part of clause 2.15.2.2 may be contrary to the Unfair Contract Terms Act 1977 as an exclusion of liability which may not satisfy the test of reasonableness.

The contractor is entitled to an extension of time if there is any delay in its receiving necessary permission or approval from a statutory body (clause 2.26.12). The contractor must have taken all practicable steps to avoid or reduce the delay. In practice that means that the contractor must have applied at the right time and not left it until there was little chance of getting the permission in time, and the delay must not be simply because the contractor defaulted in some way that an ordinary competent builder would have avoided. The contractor may also have an extension of time under clause 2.26.1 following the situations in clauses 2.15.2.1 and 2.15.2.2 discussed above.

It is clear that the contractor is not to suffer a time penalty for changes in statutory requirements or delays in receiving permissions even, probably, where the Employer's Requirements preclude payment for complying with terms in permissions relating to development control requirements under clause 2.15.2.2. The difference between predictable statutory requirements and unpredictable development control requirements is highlighted in clause 4.20, where delays in approvals for the latter, but not the former, are grounds for an application for direct loss and/or expense if incurred (clause 4.20.4). Although the draughtsman's intention may have been to separate the risks in this way, it appears that the contractor has good grounds for making application if a clause 2.15.2.1 situation arises, because clauses 2.15.2.1 and 2.15.2.2 stipulate that the amendments are to be 'treated' as if they were a change. Clause 4.20.1 expressly makes provision for changes to be grounds for direct loss and expense applications. The complexities do not end there. If delay in the receipt of any permission or approval for development control requirement purposes results in the whole or substantially the whole of the Works being delayed for a period noted in the contract particulars, the employer or the contractor may terminate the contractor's employment under the contract (clause 8.11.1.6).

The statutory requirement implications in any project entered into using this contract form deserve very careful consideration. This particularly applies to development control requirements as defined in clause 1.1. The employer who does not ensure that all such requirements are satisfied before executing the contract is taking a great risk. The employer cannot remedy the problem by simply deleting the extension of time, loss and/or expense and termination sub-clauses at the same time as inserting a statement in the Requirements that the contractor must satisfy development control requirements. If the employer followed that course of action, a failure to obtain planning permission would render the contract frustrated. That is the very reason why the termination provisions allow for termination if permission is unduly delayed, and it is noteworthy that, in such circumstances, the contractor is denied the right to claim loss and/or expense arising out of the termination.

If it becomes necessary for the contractor to comply with statutory requirements as an emergency, it must supply the minimum work and materials reasonably necessary to ensure compliance (clause 2.16.1). The contractor without delay must inform the employer of the emergency and the steps it has taken to deal with it. Unlike the situation under SBC, the contractor is not entitled to payment for compliance and, therefore, no mention is made of it. However, if the emergency compliance related to a matter which the employer had expressly stated complied with statutory requirements under clause 2.1.2, the contractor would be entitled to payment and the employer must issue an instruction under clause 2.15.2.3.

Clause 2.18 provides that the contractor must pay all fees and charges in connection with statutory requirements and the contractor must include for them in its price unless they are stated as a provisional sum in the Employer's Requirements. The contractor's obligation is not simply to pay the fees and charges, but to indemnify the employer against all liability for them. Therefore, if the contractor fails to pay a charge and the employer is obliged to pay it together with a penalty, the contractor must reimburse the employer for both charge and penalty.

6.4 The Construction (Design and Management) Regulations 2007

The 1994 Regulations were replaced in April 2007. The new Construction (Design and Management) Regulations 2007 are intended to be simplified and to combine the 1994 CDM Regulations and the Construction (Health Safety and Welfare) (CHSW) Regulations into one unit.

Article 5 assumes that the contractor will be the 'CDM Co-ordinator' under the revised Regulations. If one looks carefully at article 5, it is possible to see the word 'or', enabling the user to insert an alternative name. Article 6 defines the principal contractor as the contractor. It is not thought that these articles are sufficient in themselves to bind the contractor to the employer for the purpose of carrying out these functions and a separate contract for these services should be executed. They are certainly not sufficient if a third party is engaged as CDM co-ordinator. It also follows that termination of the contractor's employment under this contract will not automatically terminate its engagement as either CDM co-ordinator or principal contractor and express terms must be written into the ancillary contracts to achieve automatic termination.

The 'CDM Planning Period' must now be specified in the contract particulars to put into effect the requirements of Regulation 10(2)(c). The 'Health and Safety Plan' has been renamed the 'Construction Phase Plan' which, under Regulation 20, must be prepared by the principal contractor if the project is notifiable. There are sundry definitions and words which make clear that the Works must be carried out in accordance with the construction phase plan. Grounds for termination (failure to comply with the Regulations) are included in the list in both employer and contractor termination clauses (clauses 8.4.1.5 and 8.9.1.3).

Clause 3.18 provides that the employer 'shall ensure' that, where the CDM co-ordinator and principal contractor are not the contractor, they will carry out their duties in accordance with the Regulations. There are also provisions that the contractor, if, as is usual, it is the CDM co-ordinator and/or the principal contractor, will comply with the Regulations (clauses 3.18.2 and 3.18.3). The contractor must also ensure that any sub-contractor provides necessary information.

Compliance or non-compliance by the employer with clause 3.18 is a 'Relevant Event' and a 'Relevant Matter' under clauses 2.26.5 and 4.20.5, respectively. Lest the significance is missed, what this means is that the employer must ensure that, where the CDM co-ordinator and/or the principal contractor is not the contractor, they perform correctly and if they do not, or even if they do, any resultant delay or disruption will give entitlement to extension of time and loss and/or expense. This may well be a most fruitful source of claims for contractors. Every change instruction potentially carries a health and safety implication, which should be addressed under the Regulations. The Regulations impose duties on the CDM co-ordinator. Some of these duties must be carried out before work is commenced on site, which may well present difficulties where the contractor is to be the CDM co-ordinator and the contractor is not appointed until comparatively late in the process. It may be necessary to appoint a CDM co-ordinator and then replace them with the contractor on acceptance of tender. If necessary actions delay the issue of a change instruction or, once issued, delay its execution, the contractor may be able to claim.

There may be rare occasions when the Regulations do not fully apply to the Works as described in the contract. If the situation changes due to the issue of an instruction or some other cause, the employer may be faced with substantial delay as appointments of CDM co-ordinator and principal contractor are made and appropriate duties are carried out under the full Regulations.

Clause 3.19 makes provision for the situation which may arise if the employer replaces either the CDM co-ordinator or the principal contractor. The employer must immediately send written notification of the name and address of the replacement to the contractor. Where the contractor is replaced as principal contractor, it must carry out the reasonable requirements of the replacement principal contractor, but only to the extent necessary for compliance with the CDM Regulations. Significantly, the contract expressly precludes the contractor from receiving any extension of time in this regard. That is because anything the contractor is reasonably called upon to do to comply with the CDM Regulations is something which the contractor should have already done or should do to comply with clause 3.18 of the contract.

6.5 Work not forming part of the contract

The employer has the right to make contracts with persons other than the contractor to carry out work on the site. The employer is not entitled to deduct work from the contractor so as to give it to another contractor. That would be a breach of contract which is certainly not contemplated by either clause 5.1.1.1 or clause 2.6: *Vonlynn Holdings Ltd* v. *Patrick Flaherty Contracts Ltd* (1988); *AMEC Building Contracts Ltd* v. *Cadmus Investment Co Ltd* (1997). Clause 2.6 provides for two situations:

- The Employer's Requirements may provide the contractor with very full information so that it can properly carry out the work required of it under the contract. They may also note that specific work is not to form part of the contract and will be carried out by others. In such an instance, the contractor must permit the specific work.

- The Employer's Requirements may not provide the full information noted above. In that case, the employer may still employ other persons to do work not included in the contract provided that the contractor gives its consent. The contractor must not unreasonably delay or withhold its consent, but in the context of a design and build contract, it is suggested that it will be more difficult to prove that the contractor is unreasonable than under a traditional procurement system where responsibilities are already split.

Delays caused by persons engaged by the employer under clause 2.6 may give rise to an extension of time for the contractor under clause 2.26.5 or the payment of direct loss and/or expense under clause 4.20.5. The delays may be caused by persons failing to carry out the work, carrying it out slowly or simply carrying it out properly. For the employer to engage other contractors is ill advised at the best of times, but when a design and build contract is involved, it is like signing a blank cheque. After the employer has deliberately chosen a contract which puts as much responsibility as possible on the contractor's shoulders, it seems perverse to use this clause to take some of that responsibility away.

6.6 Third party rights and collateral warranties

DB has been made (some might say 'needlessly') complicated by the insertion of provisions dealing with third party rights and collateral warranties. These are dealt with in clause 7, schedule 5 and the contract particulars part 2.

6.6.1 Third party rights

Until 1999, there was a long-established principle of law that only the parties to a contract could exercise rights under that contract: *Tweddle* v. *Atkinson* (1861). Therefore, if parties A and B entered into a contract, one of whose terms is that A and B would each give third party C £500, C could not insist on receiving the money. If A or B failed to pay, there was nothing that C could do about it. If A and B entered into a contract which provided that, on a certain date, C would give each of them £500, they could not insist on receiving the payment from C. This principle was known as 'privity of contract'. It was a straightforward and easily understood principle, but one which occasionally gave rise to perceived injustice.

The matter was addressed in the Contracts (Rights of Third Parties) Act 1999 which essentially provided that parties who are not parties to a particular contract (i.e. third parties) can have rights under that contract if the contract gives that right. The contract must confer a benefit and the third party must be identified by name, class or by description. It is possible for a contract to contain a term which expressly excludes the operation of the Act, effectively reinstating the privity of contract position.

All the JCT contracts up to the 2005 series, and most other contracts, excluded the Act. DB, however, does not entirely exclude the operation of the Act. Clause 1.6 states that, except for rights of purchasers, tenants and/or a funder as set out under clause 7A and 7B, no rights are conferred on third parties. The exceptions are significant. The JCT took the decision to make use of the Act in order to avoid at least some collateral warranties. The reasoning is that if third parties can be given rights directly against the contractor through the DB contract, it saves having to organise a sheaf of warranties, especially where a multitude of purchasers or tenants are involved. Third party rights are dealt with in clause 7. Because there are only two parties to the contract and the employer is not usually called upon to give a warranty, the third party rights only affect the contractor in favour of third parties.

The mechanism provided in the contract for imparting third party rights is complicated. It remains to be seen whether it proves to be too complex for common use and, like the nomination provision in JCT 98 or the contractor's priced statement in most of the JCT contracts, is removed at a future revision. The first important point to note is that all notices given under clauses 7A–7E must be given by actual, special or recorded delivery. Special delivery is to be preferred, because next day delivery can be ensured. Rights for purchasers and tenants are covered in clause 7A and funders are covered in clause 7B.

The key to the successful operation of these clauses lies in part 2 of the contract particulars. There is a choice between completing part 2 or setting out the information on another sheet or sheets. If the information is included on other sheets, they must be clearly identified as being part of the contract. This is best done by signing, dating and fastening them to the printed form of contract. The information on separate sheets must satisfy the information requested in part 2. The names, classes or descriptions of the purchasers or tenants must be set out in section (A) of part 2, together with the part of the Works concerned and whether clause 7A or 7C (collateral warranties) is to apply. A note states that if neither 7A nor 7C is stated, 7A (third party rights) will apply as the default. Third party rights or collateral warranties can only be applied to the persons listed. Therefore, a person not listed cannot be given rights. Care should be taken that, if actual names cannot be inserted, unambiguous classes or descriptions are given.

The third party rights are set out in schedule 5, part 1 of which deals with purchasers' and tenants' rights and part 2 deals with the funder's rights. It is no surprise that the wording of these rights closely echoes the wording in standardised warranties. Variable parts of the rights must be inserted in section (B) of part 2 of the contract particulars, which deals with liabilities.

Clause 7A.1 provides that third party rights will becomes operative on receipt by the contractor of a notice from the employer stating the name of the purchaser or tenant, the nature of interest in the Works and that the rights 'vest' in that purchaser or tenant. One of the problems in giving rights under a contract to third parties is that those rights might affect, or in certain circumstances might be deemed to affect, the rights of the parties to the contract themselves. In order to avoid that situation, clause 7A.2 expressly stipulates that no purchaser's or tenant's consent is required before the employer and/or the contractor decide to terminate the contractor's employment under the contract; to amend, vary or waive any contract term; or to settle any dispute as they think appropriate. Clause 7A.3 is sensibly and necessarily inserted to prevent the employer and the contractor agreeing to change any part of clause 7A or schedule 5 without the consent of any purchaser or tenant already entitled to the rights. No doubt a term would be implied to that effect to avoid a ridiculous situation, but it is useful to have an express term to that effect.

Clause 7B deals with the funder's rights. The contract assumes there is only one funder. If there is more than one, the contract must be amended accordingly from singular to plural. In similar fashion to the provision for purchasers and tenants, clause 7B.1 provides that third party rights will become operative on receipt by the contractor of a notice from the employer stating the name of the funder and conferring the rights. Section (C) of part 2 of the contract particulars requires the insertion of the name, class or description of the funder. Failure to insert this information here or on a separate sheet will result in no rights for a funder being required from the contractor. Schedule 5, part 2 contains the funder's rights which, again for obvious reasons, closely resemble the common funder's warranty. Section (D) of part 2 of the contract particulars must be completed to indicate whether third party rights or a collateral warranty is required and if not completed, the default provision is third party rights. Almost unnoticed at the end of (D) is provision for the period of 7 days in clause 6.3 of the rights to be amended. This refers to what are generally known as 'step-in' rights and 7 days should be quite sufficient time for the funder to make up its mind whether to step into the shoes of the employer under the DB contract.

It is a mark of the importance of the funder that there is no attempt in the contract to interfere with its third party rights as is done under clause 7A.2 with purchasers and tenants. On the contrary, clause 7B.2 expressly states that neither the employer nor the contractor will agree to amend or vary the contract terms or rescind the contract; the contractor's right to terminate its employment or to treat the contract as repudiated is subject to clause 6 of part 2 of schedule 5 (the step-in rights). However, the employer and the contractor may agree to vary or waive any term and to settle disputes on appropriate terms without the funder's consent.

6.6.2 Collateral warranties

Clauses 7C to 7E deal with collateral warranties. Clauses 7C and 7D deal with contractor's warranties to purchasers and tenants, and to a funder, respectively while clause 7E deals with sub-contractors' warranties to purchasers, tenants, funder and to the employer.

Once again, the key lies in completion of part 2 of the contract particulars, which also contain important notes. If warranties rather than third party rights are required from the contractor, part 2 of the contract particulars should be completed accordingly. So far as purchasers and tenants are concerned, clause 7C provides that the employer must give notice to the contractor identifying the person and interest in the Works and require the contractor to provide a warranty within 14 days. The warranty must be in the form CWa/P&T and be completed as section (B) of part 2 of the contract particulars. Clause 7D, dealing with the funder's warranty, states that the employer may give notice requiring the contractor to enter into a warranty with the funder within 14 days on form CWa/F completed as section (D) of part 2 of the contract particulars.

Sub-contractor's warranties introduce other complexities. Section (E) of part 2 must be completed or separate sheets attached as before. The sub-contractors and any consultants employed by the contractor from whom warranties are required, must be listed. In each case, it must be stated what type of warranty is required. In the case of 7E, all the previously identified purchasers, tenants, the funder and the employer are entitled to the sub-contractor's warranties (SCWa/P&T or SCWa/F or SCWa/E). If the sub-contractor has design responsibilities, the level of professional indemnity insurance must be stated. The details entered in sections (B) and (D) will apply to the warranties.

Clause 7E provides that, within 21 days of receipt of the employer's notice listing the sub-contractor, warranty and beneficiary, the contractor must comply with the provisions in the contract documents to obtain the warranties. Importantly, the clause provides for the warranties to be obtained in a form to suit consultants and in each case amended as suggested by the relevant sub-contractor if the employer and the contractor approve. Such approval is not to be unreasonably delayed or withheld.

Chapter 7
Possession, Practical Completion and Rectification

7.1 Possession and deferment

Clause 2.3 provides that on the date of possession, possession of the site must be given to the contractor. Possession is the next best thing to ownership. A person who is in possession of something, be it a car, a television set or a building site, has a better claim to it than any other person with the exception of the actual owner. The effect of this clause is to give the contractor a licence to occupy the site for the purpose of carrying out the construction. The owner has no general power to revoke the contractor's licence, but it would be brought to an end by completion of the Works or by termination of the contractor's employment under the provisions of the contract (see Chapter 12), or if the contractor's employment or the contract itself is lawfully brought to an end by some other circumstance. If there was not an express term giving possession, a term would be implied that the contractor must have possession in sufficient time to allow it to complete by the contract completion date: *Freeman & Son* v. *Hensler* (1900).

The contractor must have possession of the whole site: *Whittal Builders* v. *Chester-Le-Street District Council* (1987). Since the contractor is in control of the site, it may exclude all other persons from the site except those persons to whom the contract expressly allows access under clause 3.1 (see section 5.4.2), or those bodies which have powers of entry under statute. If the employer fails to give possession on the due date it is a breach of contract of a fundamental nature which entitles the contractor to damages. If there were no provision for extension of time in such circumstances, time would become at large. The contractor's obligation would be simply to complete the Works within a reasonable time: *Rapid Building Group* v. *Ealing Family Housing Association* (1985). If the failure was prolonged, the contractor would have the right to treat it as a repudiation of the contract on the part of the employer. To overcome this problem, DB, in common with most JCT contracts, has a clause (2.4) which allows the employer to defer giving possession for a period which must not exceed 6 weeks. This clause applies only if so stated in the contract particulars and, therefore, it is vital that it is stated to apply or the employer will face serious consequences for any failure. A lesser period than 6 weeks may be specified in the contract particulars and, clearly, the shorter the period, the less the contractor will feel inclined to increase its tender figure. It is possible, of course, to amend the provision so that a very much longer period is specified, but the employer will be appropriately penalised in the contractor's tender. In practice, failure to give possession is a matter of either a few days or many weeks. Either the date is just missed, probably through some minor carelessness, or there is a major problem.

The precise period for insertion is something to be discussed between the employer and the employer's professional advisors.

Even where the employer has wisely stated that clause 2.4 is to apply, some consequences of deferring possession cannot be avoided. The contractor will be entitled to recover any direct loss and/or expense it has incurred under clause 4.20.5 and it will be entitled to an extension of time under clause 2.26.3. From the employer's point of view, the ability to make an extension of time for deferment of possession prevents time becoming at large and, therefore, preserves the employer's right to deduct liquidated damages for any culpable delay on the part of the contractor. There is no prescribed form for the notice of deferment and the employer need not give any reason, although it would be courteous to do so. Even if the failure to give possession exceeds 6 weeks, the contract now provides mechanisms for extension of time and recovery of loss and/or expense under clause 2.26.5 and 4.20.5 respectively, to deal with the consequences of impediment, prevention or default by the employer. However, this does not preclude the contractor from treating the breach as repudiatory in appropriate circumstances.

Having established that, under clause 2.3, the contractor retains possession until practical completion, clause 2.5 goes on to state that if the contractor gives its written consent, the employer may use or occupy the site or the Works or any part before practical completion of the Works or of any section. The employer may use or occupy for the purpose of storage of goods or otherwise – which is fairly broadly drafted. The reason for the clause appears to be to enable the employer to store goods on the Works without the necessity for operating the partial possession clauses (clauses 2.30–2.34, but it is also possible to use it to permit the employer to occupy the whole site: *Skanska Construction (Regions) Ltd* v. *Anglo-Amsterdam Corporation Ltd* (2002)). There is a procedure which must be observed before the contractor may give consent. Either the contractor or the employer must notify the insurers under the appropriate insurance provision (options A, B or C). If the insurers confirm that the use or occupation will not prejudice the insurance, the contractor may not withhold permission without good reason. In practice, finding reasonable grounds for withholding consent should not present too much of a problem.

Clause 2.5.2 stipulates that if the insurers have made a condition that an additional premium is required under option A, the contractor must notify the employer. The employer, in turn, must state whether use or occupation under this clause is still required and if so, the additional premium must be added to the contract sum and if the employer requires a receipt, the contractor must provide one.

7.2 Progress

Clause 2.3 stipulates that the contractor must regularly and diligently proceed with the construction of the Works and complete them 'on or before' the completion date. The precise meaning of 'regularly and diligently' has been the subject of some discussion. In *London Borough of Hounslow* v. *Twickenham Garden Development Ltd* (1970), the judge considered the meaning for some time and concluded: 'At present, all that I can say is that I remain somewhat uncertain as to the concept enshrined in these words'. Perhaps more helpful are the observations of the court in *Greater London Council* v. *Cleveland Bridge & Engineering Co Ltd* (1986), which were approved by the Court of Appeal. The court was considering the meaning of 'due diligence and expedition' which was not actually included as an obligation of the contractor, but only in the negative way that failure to work with due diligence was grounds for termination:

> '. . . I would have held without hesitation, that due diligence and expedition must be interpreted in the light of the other obligations as to time in the contract. That seems to me to follow from the construction of the contract as a whole and from the considerations I have already mentioned . . . If there had been a term as to due diligence, I consider that it would have been, when spelt out in full, an obligation on the contractors to execute the works with such diligence and expedition as were reasonably required in order to meet the key dates and completion date in the contract.'

This has to be considered together with the general principle that a contractor is entitled to plan and carry out the work to suit itself provided that there is no provision to the contrary and that it meets the completion date: *Wells* v. *Army and Navy Co-operative Society* (1902). The employer may impose restrictions in the Requirements or in a change instruction referring to clause 5.1.2. The principle to be derived from those cases seems to be that it is difficult to successfully contend that a contractor is failing to proceed regularly and diligently provided it is doing some work on the site and provided that it can meet the completion date. However, the Court of Appeal in *West Faulkner* v. *London Borough of Newham* (1995) have helpfully defined 'regularly and diligently' in terms which make it easier for the employer to allege such failure:

> 'What particularly is supplied by the word "regularly" is not least a requirement to attend for work on a regular daily basis with sufficient in the way of men, materials and plant to have the physical capacity to progress the works substantially in accordance with the contractual obligations.
>
> What in particular the word "diligent" contributes to the concept is the need to apply that physical capacity industriously and efficiently towards the same end.
>
> Taken together the obligation upon the contractor is essentially to proceed continuously, industriously and efficiently with appropriate physical resources so as to progress the works steadily towards completion substantially in accordance with the contractual requirements as to time, sequence and quality of work.'

Where the Works are to be completed in sections, certain key dates are indicated which must be met. Provided such key dates are in the contract, they will be binding on both employer and contractor and the contractor's progress can be measured accordingly.

Under normal circumstances, the contractor's programme is not binding on either employer or contractor. It is very unusual for it to be made a contract document. It is quite clear, from clause 2.3, that the contractor must finish by the completion date, but it may finish before such date. The point was emphasised so far as the JCT Standard Form 1963 was concerned in *Glenlion Construction Ltd* v. *The Guiness Trust* (1987). By analogy, it is thought that the other holding in that case, that the architect is not obliged to produce information to suit the contractor's shortened work period, applies to decisions and approvals which the employer may have to give under this form.

The contractor's right to complete before the completion date can be a source of embarrassment to an employer who has scheduled finances in fairly equal amounts to be paid throughout the contract period. An employer who depends on income to fund payments may be driven into overdraft. It is difficult to avoid this situation, but it can be alleviated by the simple expedient of amending clause 2.3 to omit the words 'or before' after 'the same on', so that the contractor's obligation becomes to complete 'on the Completion Date'. If this cause of action is taken, the contractor's attention should be particularly drawn to the change: *J Spurling Ltd* v. *Bradshaw* (1956); *Interfoto Picture Library* v. *Stiletto Visual Programmes* (1988). It is, of course, still open to the contractor to complete most of the work and leave very little to finish in the last part of the contract period. To properly regulate payment is possible, but it would require much more severe re-drafting.

Under the provisions of clause 3.10, the employer has the power to postpone any design or construction work which the contractor is to carry out under the terms of the contract (see section 5.3.2). Note the clause no longer expressly refers to 'design'; that is included as part of the work the contractor is obliged to carry out.

7.3 Practical completion

Practical completion is dealt with in clause 2.27. There is a significant difference between practical completion in this contract and in SBC. In SBC, a certificate is to be issued by the architect when, in the architect's opinion, practical completion has been achieved. DB merely provides that the employer must give the contractor a written statement when the Works have reached practical completion and when the contractor has complied sufficiently with clauses 2.37 and 3.18.5 – in effect when the contractor has provided the employer with as-built drawings and the health and safety file required by the CDM Regulations. Under SBC it is a matter for the architect's opinion, while under DB it is a matter of fact. Since the employer has merely to state fact, rather than certify opinion as an independent professional, the statement is not thought to be binding until arbitration: *J F Finnegan Ltd* v. *Ford Seller Morris Developments Ltd (No. 1)* (1991). The contract, however, states that for all purposes of the contract, practical completion is deemed to have taken place on the date named in the statement.

The introduction of a requirement to comply with clauses 2.37 and 3.18.5 has introduced some ambiguity into this clause. What it amounts to is that for practical completion to be 'deemed' to have taken place for all the purposes of the contract, two criteria must be satisfied:

- The Works must have reached practical completion; and
- The contractor must have complied sufficiently with clauses 2.37 and 3.18.5.

One of these criteria is clearly practical completion of the physical Works. Just as clearly, one criterion is not practical completion. Yet the clause makes clear that practical completion will not be deemed to have taken place until both are satisfied. The inclusion of a 'deeming' provision has the effect that although both parties recognise that practical completion has not taken place on that day (indeed it may have taken place in a physical sense some time earlier), they both behave as though it took place on the date stated: *Re: Coslett (Contractors) Ltd, Clark, Administrator of Coslett (Contractors) Ltd (in Administration)* v. *Mid Glamorgan County Council* (1997). This may have repercussions on the liquidated damages position (see section 8.4).

The contract does not define what is meant by practical completion of the Works and there is conflicting case law. A sensible approach seems to be the one taken in *H W Neville (Sunblest) Ltd* v. *Wm Press & Sons Ltd* (1981) where it was held that at practical completion there should be no defects apparent although there might be trifling items outstanding. The idea was explained further in *Emson Eastern Ltd (in Receivership)* v. *E M E Developments Ltd* (1991):

> 'I think that the most important background fact which I should keep in mind is that building construction is not like the manufacture of goods in a factory. The size of the project, site conditions, use of many materials and employment of many types of operatives makes it virtually impossible to achieve the same degree of perfection as can a manufacturer. It must be a rare new building in which every screw and every brush of paint is absolutely correct.'

A practical test was suggested by Lord Justice Salmon in the Court of Appeal considering the 1963 standard form of contract, and it was not disapproved on appeal to the House of Lords, in *Westminster Corporation* v. *J Jarvis & Sons* (1968):

> 'The obligation upon the contractors under clause 21 to complete the works by the date fixed for completion must, in my view, be an obligation to complete the works in the sense in which the words "practically completed" and "practical completion" are used in clauses 15 and 16 of the contract. I take these words to mean completion for all practical purposes, that is to say, for the purpose of allowing the employer to take possession of the works and use them as intended. If completion in clause 21 meant completion down to the last detail, however trivial and unimportant, then clause 22 would be a penalty clause and as such unenforceable.'

When the case went to the House of Lords, they said:

> 'The defects liability period is provided in order to enable defects not apparent at the date of practical completion to be remedied. If they had been apparent, no such certificate would have been issued.'

In *P & M Kaye Ltd* v. *Hosier & Dickinson Ltd* (1972), the court held that the architect could withhold the certificate until everything except trifling defects were rectified.

It is clear that practical completion falls short of completion in every particular and it seems that it is achieved when there are no significant visible defects and there are only minor things left to complete. Minor things are probably those the rectification of which will not inconvenience the employer.

If the contractor maintains that practical completion has taken place and the employer declines to give a statement to that effect, the contractor may refer the question to adjudication under clause 9.2. In practice, the parties normally carry out a joint inspection. The contractor will be anxious to secure the statement that the Works have reached practical completion, because some very important consequences flow from it:

- The contractor's liability for insurance under schedule 3, option A ends.
- Liability for liquidated damages under clause 2.29 ends.
- The employer's right to deduct full retention ends; half the retention becomes due for release (clause 4.17.3).
- The 3-month period begins during which the contractor must submit the final account and final statement (clause 4.12.1).
- The period for final review of extension of time begins (clause 2.25.5).
- The rectification period begins (clause 2.35).

If the Works are divided into sections, the employer must issue a separate section completion statement on practical completion of each section. When issuing the section completion statement for the last section, the employer must issue a practical completion statement for the whole Works at the same time in order to place beyond doubt that practical completion of the Works has been achieved. That is because the sum of all the sections, as defined in the contract, may not always amount to the Works and it is the 'Works' that the contractor undertakes to complete.

7.4 *Partial possession*

The contract makes provision for partial possession by the employer in clauses 2.30–2.34. This is not intended for use as a means of achieving sectional completion, because the contractor cannot be compelled to complete in predetermined parts; provision is made in the contract particulars for sections. Clause 2.30 provides that the employer may take possession of any part of the Works before practical completion if the contractor's consent has been obtained. This power is said to be in spite of anything which may be expressed or implied elsewhere in the contract. Therefore, a conflict is avoided between this clause and, for example, clause 2.3 which effectively gives complete possession to the contractor until completion. The contractor's consent must not be delayed or withheld unreasonably. In contrast to the provisions for practical completion, it is the contractor who must issue the employer with a written statement. The statement must identify the part or parts of the Works taken into possession and the date possession was taken. The clause proceeds to refer to the matters so identified as the 'Relevant Part' and the 'Relevant Date' respectively. The contractor is not expressly required to estimate the value of the relevant part, but the making of an estimate is implied in clause 2.34.

Certain consequences follow after the contractor issues the written statement. Each of them is beneficial to the contractor and it is seldom that it will refuse to allow partial possession. They broadly echo the consequences of practical completion. Indeed, it is almost as if practical completion has been reached as regards the relevant part. The consequences are:

- The rectification period commences and half the retention sum is released in respect of the relevant part (clause 2.31).

- After defects, notified under clause 2.35 in respect of the relevant part, have been made good, the employer must issue a notice to that effect (clause 2.32).
- From the relevant date, the obligation to insure the part taken into possession, whether it be by the contractor under option A or by the employer under option B or C, ends. If the employer is insuring under option C, the relevant part must be included in the insurance of paragraph C.1, existing structures (clause 2.33).
- If the contractor becomes liable to pay liquidated damages after the relevant date, the amount of damages payable is to be reduced pro rata to the value of the relevant part as a proportion of the contract sum. This simple clause replaces a clause of awesome complexity in WCD 98 (clause 2.34).

There is a school of thought which contends that the operation of this provision may be sufficient to change liquidated damages into a penalty on the basis that a genuine pre-estimate of loss in respect of a building as a whole may not be a genuine pre-estimate of loss when reduced simply on a pro rata basis. A relatively insignificant part of the building in terms of straightforward construction cost may be the most valuable in terms of use by the employer. Once that part is taken into possession, the reduction in liquidated damages does not properly reflect reality. Against such arguments may be set the thought that liquidated damages are commonly rather less than a genuine pre-estimate of loss and the courts tend to take a pragmatic view of their operation: *Philips Hong Kong Ltd* v. *Attorney General of Hong Kong* (1993).

7.5 Rectification period

Respective liabilities during and after the rectification period are much misunderstood. The period is to be stated in the contract particulars. It is important that a period is inserted, because failure to name a period will mean that the period will be 6 months from the date of practical completion. That may be perfectly satisfactory, but it is becoming common for employers to insert 12 months as a more appropriate period, because it exposes the building to the full range of seasonal differences. Any defect in the Works is a breach of contract on the part of the contractor. The idea of the rectification period is to allow a reasonable period for defects to become apparent and for the contractor to rectify them. This saves the employer the time and effort of rectification by another and taking legal action for the cost. From the contractor's point of view, it can rectify its own defects more cheaply than another contractor who is a stranger to the Works. If it were not for this clause, the contractor would have no right or duty to enter the site after practical completion. Its licence would have expired (see clause 2.3 and section 7.1).

The employer has 14 days after the end of the rectification period in which to deliver a schedule of defects to the contractor and the contractor has a 'reasonable time after receipt' in which to make good the defects. What is a reasonable time will depend on the number and the type of defects.

Contrary to the commonly held view, the contractor's liability for defects does not end when the rectification period ends. What ends is simply its right to rectify the defects. It will be seen that even this right is severely circumscribed.

The rectification period is commonly referred to as the 'maintenance period'. Some other forms of contract, such as ACA 3 or GC/Works/1(1998) adopt that terminology. However, it is best avoided because it implies the considerably more onerous duty to keep in pristine condition, even though the contracts, where the term is used, restrict the effect to the equivalent of rectification in JCT contracts.

The defects which the contractor is to make good are spelled out in detail. They are 'any defects, shrinkages or other faults'. Certain criteria must be satisfied. The defects must appear within the rectification period and they must be due to the contractor's failure to comply with its obligations under the contract. The phrase 'defects, shrinkages and other faults', which might be taken to be extremely wide, is somewhat less so and 'other faults' is to be interpreted *ejusdem generis*. Thus, the faults must be of the same class as defects and shrinkages. The contractor's failure to comply with its obligations under this clause may be referable to design or construction. A defect which stems from some other cause, such as some inadequacy in the Employer's Requirements, is not covered by this clause. The fact that the contractor also has design obligations under this contract should eliminate many of the arguments about defects. For example, it is not open to the contractor to argue that timber shrinkage is due to inadequate specification if the specification is part of its Proposals. A strict reading of the clause suggests that defects which are outstanding at practical completion may not be considered as falling within this clause. This reinforces the definition of practical completion as being the point at which there are no apparent defects. However, to avoid absurdity it should be interpreted as including any defects which are apparent at practical completion: *William Tomkinson & Sons Ltd* v. *The Parochial Church Council of St Michael and Others* (1990). The employer must specify the defects which satisfy the criteria set out in the clause and they must be delivered to the contractor as an instruction no later than 14 days after the end of the period. The contractor's obligation is then to make good the defects specified within a reasonable time at its own cost. What is reasonable will depend on the number and kind of defects in the instruction. There is an important proviso that the employer may instruct that some or all of the defects are not to be made good and an appropriate deduction is to be made from the contract sum.

There is no definition of an 'appropriate deduction'. It is often contended by the employer that it is the cost to the employer of having the defects made good by others. The contractor, understandably, will argue that the deduction should be the cost which the contractor would have expended on making good. It seems unlikely that the employer's contention is correct if the contractor is willing to make good the defects. Probably the clue lies in the fact that if the employer instructs the contractor not to make good certain defects, it is not, or not necessarily, because the contractor has defaulted, but because the employer has chosen to exercise a right under the contract and in any event, the employer has an ordinary duty to mitigate loss. This may be because the employer has lost all faith in the contractor's ability to rectify the defects, but it may be because the employer is prepared to put up with various minor imperfections to avoid disturbance to the office, factory or home. The appropriate deduction is thought to be the cost which the contractor would have incurred in rectifying the defect. This view is strengthened by the observations of Mr Justice Stannard in the *Tomkinson* case noted above when considering a very similar point:

> '... but the true measure of damages which governs this aspect of the case is not the church's outlay in remedying the damage, but the cost which the contractors would have incurred in remedying it if they had been required to do so.'

There seems little doubt that if the contractor refuses to make good the defects on the schedule, or if it does not expressly refuse but simply does not make good, the employer would be able to instruct the contractor not to make good in accordance with clause 2.35 and the appropriate deduction in such an instance would be the cost to the employer of engaging another contractor to make good. If the employer were to forget to issue the schedule until some time after the 14 days had expired, the defects would still amount to breaches of contract on the part of the contractor, but it could not be compelled to return to make good. The employer could engage others to make good, but the amount to be deducted from the contract sum would be what it would have cost the contractor to make good if the employer had notified it of the defects at the correct time: *Pearce & High* v. *John P Baxter & Mrs A Baxter* (1999). It has been held in a Scottish case that an employer under the JCT Design and Build Contract is not entitled to recover the cost of hiring an architect for the purpose of inspecting the Works for defects after practical completion. Although the defects are breaches of contract, the contract prescribes the remedy and, if the contractor remedies the defects, the architect's fees are not recoverable as damages: *Michael Johnston* v. *W H Brown Construction (Dundee) Ltd* (2000).

Important powers are conferred upon the employer by clause 2.35.2. The employer may issue instructions for the making good of any defect which satisfies the same criteria as laid down in clause 2.35. These instructions may be issued whenever the employer considers it necessary to do so. It is clear that this power is quite separate from the requirement to prepare a list of defects at the end of the rectification period. Within a reasonable time, the contractor shall comply at no cost to the employer unless the employer instructs otherwise, when an appropriate deduction from the contract sum must be made. No such instructions can be issued after the earliest of either the delivery of the defects schedule or the expiry of 14 days following the end of the rectification period.

When the contractor has made good all the defects notified by the employer under clause 2.35, the employer must issue a notice to that effect. It must not be unreasonably delayed or withheld, because it affects the final account and final statement (see section 10.6). The contract states that making good defects is to be deemed to have taken place on the date named in the notice 'for the purposes of this Contract' (clause 2.36). This is presumably to make clear that the notice is for no other purpose and particularly not for a purpose associated with any other contract, for example a tenancy agreement.

Defects which appear after the end of the rectification period are still the liability of the contractor. Although it can no longer demand the opportunity to make good as a contractual right, the principle of mitigation of loss will often mean that an employer will invite the contractor to do so as the cheapest possible solution for all concerned. The contractor's liability for such latent defects will be governed by the Limitation Act 1980 subject to whatever may be the conclusive effect of the final account and final statement (see section 10.6).

Chapter 8
Extension of Time

8.1 Principles

Under the general law and in the absence of any contractual provision empowering the award of an extension of time, the contractor is bound to complete the Works by the date agreed, unless it is prevented from so doing by some action or inaction of the employer. The position was neatly put by Lord Fraser of Tullybelton in *Percy Bilton Ltd* v. *Greater London Council* (1982):

> 'The general rule is that the main contractor is bound to complete the work by the date for completion stated in the contract. If he fails to do so, he will be liable for liquidated damages to the employer . . . That is subject to the exception that the employer is not entitled to liquidated damages if by his acts or omissions he has prevented the main contractor from completing his work by completion date . . . These general rules may be amended by the express terms of the contract.'

This general rule is amended in DB by clauses 2.23–2.26 which expressly confer upon the employer the power to extend the contract period for specific reasons. If there were no such clauses and the employer prevented completion by the due date, or if the employer fails to give an extension of time as provided for under the clause, the contractor's obligation to complete by the contract date for completion is removed and its obligation is merely to complete within a reasonable time: *Wells* v. *Army & Navy Co-operative Society Ltd* (1902). Even if the contractor is subsequently delayed through its own fault, the employer cannot then deduct liquidated damages, because there is no fixed date from which the damages can be calculated: *Miller* v. *London County Council* (1934). If the employer's right to recover liquidated damages is to be kept alive, it is essential that the extension of time clause is operated properly and promptly, at least where the grounds for the award are the fault or responsibility of the employer. In that sense, the extension of time clause is for the benefit of the employer. Of course, it also benefits the contractor when it provides for an extension of time on grounds which are outside the employer's control and for which the contractor otherwise would have to take the risk. 'Exceptionally adverse weather conditions' are one such ground.

In the New Zealand case of *Fernbrook Trading Co Ltd* v. *Taggart* (1979), Mr Justice Roper took the view that, under the normal extension of time clause, a retrospective extension of time is only valid in two circumstances:

> '(1) Where the cause of delay lies beyond the employer and particularly where its duration is uncertain . . . although even here it would be a reasonable inference to draw

from the normal extension clause that the extension should be given a reasonable time after the factors which will govern the engineer's discretion have been established.

(2) Where there are multiple causes of delay there may be no alternative but to leave the final decisions until just before the issue of the final certificate.'

In another New Zealand case, *New Zealand Structures & Investments Ltd* v. *McKenzie* (1979), a different judge took the view that under the normal extension of time clause the certifier can grant an extension of time right up until the time of becoming *functus officio*, i.e. devoid of powers, which in most cases will be on the issue of the final certificate. The court said:

> 'In a major contract it is virtually impossible to gauge the effect of any one cause of delay while it is still proceeding, let alone assess the consequences of concurrent or overlapping causes. Finally, any need to have a prompt decision loses some force as a factor in interpreting such a clause, when one considers the normal review and arbitration procedures . . .'.

This, in general, is a realistic approach.

It is commonly assumed by contractors that they must first secure an extension of time before they are entitled to claim direct loss and/or expense. Under traditional contracts, architects and quantity surveyors are often of the same mind. The sequence is often that the contractor first obtains an extension of time for various reasons. It then applies for direct loss and/or expense and the quantity surveyor is instructed to value the 'cost related' extensions at a figure per week extracted from the preliminaries to the bills of quantities or from actual costs. Many claims are settled quite amicably on this basis. However, it is not strictly correct. The contractor may have an entitlement to an extension of time but not to loss and expense, or it may be entitled to loss and expense but not to extension of time, or it may be entitled to extension of time and loss and expense. The obtaining of an extension of time is not a precondition to the recovery of loss and expense: *H Fairweather & Co Ltd* v. *London Borough of Wandsworth* (1987). The misconceptions almost certainly arise from the fact that some of the grounds for extension of time are reflected almost word for word in clause 4.20 (loss and expense). The situation was explained in *Henry Boot Construction Ltd* v. *Central Lancashire New Town Development Corporation* (1980) by Judge Edgar Fay QC, speaking about the somewhat similar clauses in the 1963 JCT Standard Form:

> 'The broad scheme of the provisions is plain. There are cases where the loss should be shared, and there are cases where it should be wholly borne by the employer. There are also those cases which do not fall within either of these conditions and which are the fault of the contractor, where the loss of both parties is wholly borne by the contractor. But in the cases where the fault is not of the contractor the scheme clearly is that in certain cases the loss is to be shared: the loss lies where it falls. But in other cases the employer has to compensate the contractor in respect of the delay, and that category, where the employer has to compensate the contractor, should, one would think, clearly be composed of cases where there is fault upon the employer or fault for which the employer can be said to bear some responsibility.'

The effect of any delay is to be considered in relation to what is actually happening on site at the time the delaying factors operate, not in relation to what should have been happening on site by reference to any programme: *Walter Lawrence & Son Ltd* v. *Commercial Union Properties (UK) Ltd* (1984).

8.2 Contract procedure

Clauses 2.23–2.26 closely follow the extension of time provisions in SBC. There are some differences. The contractor must give notice in writing to the employer every time it becomes reasonably apparent that progress is being or is likely to be delayed. That is made plain in clause 2.24.1. Common sense suggests that it is the contractor to whom it is to be reasonably apparent, because until it knows, it cannot act. The written notice must give the circumstances of the delay and, if the contractor considers that it is entitled to an extension of time, it must state the relevant event (clause 2.26) which covers the situation. The contractor must not simply notify delays for which it expects the contract period to be extended, but any delay to progress. For example, if an important piece of earth-moving machinery breaks down and takes 3 days to fix or replace, the contractor must report the fact to the employer. The idea is that the employer is kept fully informed of all factors which might result in the project completion being delayed, so that the employer can take action, for example, by issuing instructions to replace floor tiles with a more readily available product to make up for the employer's own delays, and to carefully monitor the results of the contractor's delays. If the contractor fails to give written notice of every delay, it is a breach of contract which the employer is entitled to take into account when considering an extension of time: *London Borough of Merton* v. *Stanley Hugh Leach Ltd* (1985) – which also made clear that the giving of the notice is not a precondition to the award of an extension of time.

If the delay notified by the contractor is not identified as a relevant event, its duty ends there until another delay occurs. If, however, the delay is a relevant event, the contractor must give further information in the notice or as soon as possible afterwards. The contractor must take each separate relevant event and estimate the effect on other items of work and the effect on the completion date. It may be that a delay, although it is a relevant event, has no effect on the completion date, possibly because it is not on the critical path of the project. If that is the case, the contractor must so state. Thus, the contractor may refer to three different relevant events and give the effects as 2 days, 1 week, 3 weeks, respectively. That does not mean that the cumulative result will be equal to a simple aggregate, i.e. 4 weeks and 2 days. Some delays may have concurrent effects. It is quite difficult to isolate the effects in this way and many contractors call in aid the computer to perform this chore. There are several excellent software packages which will allow the contractor to input its programme as a network and separately introduce delays, and this approach has judicial approval: *John Barker Ltd* v. *London Portman Hotels Ltd* (1996); *Balfour Beatty Construction Ltd* v. *London Borough of Lambeth* (2002). However difficult the task, the contractor must do its best. In cases where the situation is changing, the contractor must keep the information updated.

The employer's duties are set out in clause 2.25.2. The employer has 12 weeks in which to fix a new completion date. The 12 weeks is measured from receipt of the contractor's notice under clause 2.24.1 and particulars required by the employer. If there is less than 12 weeks between the receipt of the information and the date for completion, the employer must endeavour to make a decision before the date for completion. This may not be easy if the information is not available until only 1 week before completion date. In such circumstances, the employer's obligation is to give a decision as soon as reasonably practicable, but not to exceed 12 weeks.

For purposes of giving extensions of time, the total timescale is divided into two periods, like the position under SBC. The first period runs from the date of possession until the date of practical completion. The second period runs from the date of practical completion until 12 weeks after practical completion. Therefore, the employer must respond to all notices from the contractor (provided that all the required particulars have been submitted) as soon as reasonably practicable, but no later than 12 weeks. From the contractor's point of view, this is a big improvement on the position under WCD 98 where it appeared that information provided too close to the completion date or issued after that date could be left to be decided after practical completion.

What constitutes required particulars amounts to the information noted under clause 2.24.2 and any further information required by the employer under clause 2.24.3, i.e. at least in respect of each relevant event, the causes and the effects on other work, together with an estimate of the effect on completion date. If the delay is continuing, it seems that the employer is not obliged to fix a new date until after the full details of the effects of the delay have been obtained. That date, therefore, should be capable of identification without too much difficulty.

The criteria which the employer must bring to the fixing of a new completion date are threefold:

- The date for completion must be fair and reasonable.
- The event must be a relevant event (i.e. one of those listed under clause 2.26).
- The relevant event must be likely to result in the completion of the Works being delayed.

In *John Barker Ltd*, the judge's view was that extensions of time should be properly calculated and not made on the whim of the architect. Particularly, the contractor's own delays should not be taken into account. We know of many architects who decide on extensions of time by taking the total time taken by the contractor to complete and then deducting what the architect feels to be the time for which the contractor is responsible. This kind of negative approach may have certain attractions, but it is most definitely not what the contract requires. However, in deciding whether a relevant event has caused delay, the employer is entitled to consider the impact of other events on progress and completion: *Henry Boot Construction (UK) Ltd* v. *Malmaison Hotel (Manchester) Ltd* (1999).

It is possible that the Works are being substantially delayed by circumstances which can be shown to fall under the head of one or more relevant events, but if it does not appear that the Works will be delayed beyond the date for completion set in the contract (or any previously extended date), the contractor is not entitled to any extension of time: *Royal Brompton Hospital NHS Trust* v. *Hammond and Others (No.7)* (2001). The fixing of an extension of time by the employer is quite a difficult business. Since the employer or agent is not under any duty to act fairly between the parties other than the contract duty to set a new date which is fair and reasonable, the employer's decision under this clause seems not in any sense to be binding unless the parties allow it be binding.

The employer, however, has got strict obligations so far as the notice fixing a new date is concerned. Apart from setting a new date, the employer must allocate the appropriate extension of time to each of the relevant events, or reduction in time attributed to each relevant omission. The allocation of extension of time to the relevant events is not good news for the employer. There was no such provision in WCD 98. The contractor will usually be anxious to see a precise apportionment, because it will then attempt to use the 'cost related' relevant events to claim loss and/or expense.

The employer is entitled to take omission of work into account. Indeed, work may be omitted specifically to avoid making a lengthy extension of time. In practice, the truth is that once the construction process is under way, omission of work is unlikely to reduce the contract period to any significant extent unless the omission is itself significant. Clause 2.25.4 makes clear that after the employer has carried out the procedure once, subsequent omissions may be taken into account to the extent of reducing a previously awarded extension of time. It seems that the employer may so act without the action being triggered by a notice from the contractor. For example, the contractor may submit a notice of delay following which the employer may fix a new date of 12 May instead of 15 April. Subsequently, the employer may issue a change instruction which has the effect of reducing the amount of work required of the contractor. The employer may then fix a new date for completion which is earlier than the 12 May so as to allow for the reduction in work.

Clause 2.25.6.4 states that on no account can a pre-agreed adjustment, referred to in clause 5.2, be changed unless the work is omitted. Although clause 2.23.2 defined 'Pre-agreed Adjustment' as the fixing of a revised completion date in respect of a change or other work referred to in clause 5.2, clause 5.2 does not mention a pre-agreed adjustment. It essentially refers to the valuation of changes. Moreover, there appears to be no express power for the employer and the contractor to pre-agree an adjustment. Of course, as parties to the contract, the employer and the contractor can agree whatever they wish. Presumably it is the possibility of this ad hoc type of agreement to which clause 2.23.2 refers and intends to put on a proper basis. It would have been helpful if the intention had been made clear. Clause 2.25.6.3 stipulates that the contractor has no power to fix a date earlier than the date for completion noted in the contract particulars.

The employer is obliged to review the extension of time after the date of practical completion. If practical completion occurs after the date for completion in the contract or as extended, the employer may carry out the review after the date for completion has passed. A strict reading of clause 2.25.5 suggests that the employer may carry out the review only once, whether it is immediately after the date for completion or just before the expiry of the appropriate period after practical completion. This is a valuable power for the employer and it can prevent time becoming at large. In carrying out the review, the employer may consider omissions of work. The employer is not tied to any previous decision or notification by the contractor and may act freely except that delays must be considered in relation to relevant events.

The employer has three options:

- To fix a completion date which is later than the completion date already fixed.
- To fix a completion date which is earlier than the completion date having regard to instructions requiring an omission issued after the last date on which time was extended.
- To confirm the completion date previously fixed.

The employer must act before the expiry of 12 weeks after the date of practical completion. It has been said that this period is merely directory, not mandatory (*Temloc Ltd* v. *Errill Properties Ltd* (1987)) and that the employer may take rather longer if desired. The views expressed in this case should be treated with care. The Court of Appeal was considering a situation where the employer was attempting to use to its own advantage the architect's failure to act. The court clearly applied the principle that a party should not be allowed to profit through its own breach (*Alghussein Establishment* v. *Eton College* (1988)) and interpreted the provision *contra proferentem* against the employer. Where time periods are specifically set out in the contract, it is always prudent to adhere to them. The 12-week deadline has since been confirmed by another court: *Cantrell* v. *Wright & Fuller Ltd* (2003).

There is an important proviso in clause 2.25.6.1. The contractor must constantly use its best endeavours to prevent delay however caused, and it must do everything reasonably required to the satisfaction of the employer to proceed with the Works (clause 2.25.6.2). The employer, when making an extension of time, is entitled to take into account the extent to which the contractor has complied with these requirements. The proviso does not empower the employer to require acceleration of the Works. If the parties agree that acceleration is possible and advisable, it must be the subject of a separate agreement. If they should embark on this course of action, however, the effect on other contract provisions must be considered. Among other things, clauses 2.25, 2.27, 2.35 and 4.19 may be affected and expert advice is required. Clause 2.25.6.1 appears to be nothing more than a duty to continue to work regularly and diligently. In the Australian case *Victor Stanley Hawkins* v. *Pender Bros Pty Ltd* (1994), it was defined as doing everything prudent and reasonable to achieve an objective. If the employer requires some action to be taken which does not involve the contractor in extra expenditure, but which may save some time, the contractor is obliged to comply.

8.3 Relevant events

The grounds which entitle the contractor to an extension of time are termed 'relevant events'. The employer may only fix a new date for completion if satisfied that the delay falls squarely within one of the relevant events under clause 2.26. The grounds comprise two categories – events which are attributable to the employer, and events which are attributable to neither employer nor contractor – and they are listed in that order in the contract. Events falling into the first group are to be found under clauses 2.26.1–2.26.5. Events falling into the second group are under clauses 2.26.6–2.26.13. The following descriptions are simplified and are not quotations from the contract, to which reference should be made for the full wording.

Clause 2.26.1 Changes or matters treated as changes

It should be noted that this event does not simply include changes instructed by the employer under clause 3.9, but also any clause which has the effect of treating something as a change, for example, clause 2.15.2.1.

Large numbers of small instructions may have effects which are quite out of proportion to their value compared to the total value of the Works. The practice of arriving at an 'appropriate' extension by simple proportioning of these values to times is quite mistaken. For example:

$$\frac{\text{Value of instruction: £50,000}}{\text{Contract sum: £500,000}} = \frac{\text{Extension of time: } x}{\text{Contract period: 10 months}}$$

Therefore the value of $x = 1$ month.

If there were 150 instructions to carry out all kinds of extra work, the delay might actually be much greater than 1 month. On the other hand, if there was only one instruction involving the pouring of many extra cubic metres of mass concrete, the delay might be very little. Indeed, it is possible that a large number of instructions, some to add, some to omit and others simply to change the work, may result in virtually no change in the contract sum, but the delaying effect may be severe nonetheless. There are no firm rules and each instruction should be separately evaluated, not only in terms of money but also in terms of delay. The disorganising effect of a multitude of instructions should also not be overlooked.

Clause 2.26.2 Employer's instruction under clauses 2.13, 3.10, 3.11 and 3.16 or regarding opening up and testing under clauses 3.12 and 3.13.3

Clause 2.13 refers to divergencies in or between the contract documents, instructions and documents issued under clause 2.8. The employer must issue instructions and if compliance results in a likely delay to completion, an extension of time is indicated. In some cases, prompt instructions from the employer will result in virtually no delay or, at worst, a delay of just a short time.

Clause 3.10 refers to the postponement of design or construction (see section 5.3). A postponement of 3 weeks in the execution of the Works midway through the project will certainly give rise to more than 3 weeks' delay to the completion of the Works as the contractor has to carry out certain procedures before stopping (for example, make the Works safe) and to gear the workforce to start again. The effect of partial postponement will depend on the position of the postponed work in the network.

An instruction for the expenditure of a provisional sum poses another kind of problem. A provisional sum can be included only in the Employer's Requirements (clause 3.11). Commonly, the information given about the subject of the provisional sum is sketchy. In many instances, the contractor cannot make any realistic provision in its programme and it is not obliged to guess any likely time. In such cases, the issue of an instruction may entitle the contractor to an extension of time which bears no relation to the difference between the provisional sum and the actual cost.

Clause 3.16 refers to the discovery of antiquities and the issuing of instructions to deal with them.

Opening up and testing of work is dealt with in clauses 3.12 and 3.13.3. The contractor is entitled to an extension of time in both instances unless the work is not in accordance with the contract.

Clause 2.26.3 Deferment of possession

If clause 2.4 is stated in the contract particulars to apply, the exercise by the employer of the power to defer possession for up to 6 weeks will inevitably attract an extension of time. The period of extension may, but will not necessarily, equal the period of deferment. It may be longer to allow for the remobilisation of labour and materials. This clause must be interpreted strictly. The employer has no power to extend time under this clause on failure to give possession if the deferment clause is not stated to apply or if the employer, having got the right, has not formally exercised it.

Clause 2.26.4 Suspension of contractor's obligations

If the contractor properly exercises its right under clause 4.11, it is entitled to an appropriate extension of time. What is appropriate is unlikely to be merely the length of the suspension. When the contractor receives payment in full, it will need time to get back to full production on site, depending on the stage the project has reached.

Clause 2.26.5 Employer impediment, prevention or default

This clause acts as a catch-all clause for any events which are the responsibility of the employer or the employer's agent. It is broad enough to encompass breaches of contract. If any default of the contractor contributed to the event, the contractor's entitlement to extension of time is to be reduced accordingly. The inclusion of this clause has made it unnecessary to include relevant events dealing expressly with late instructions or decisions from the employer, the carrying out of work which does not form part of the contract, the failure to give access to the site and the employer's compliance or non-compliance with the CDM Regulations.

Clause 2.26.6 Work by statutory undertaker

This clause applies only when the work is being carried out in relation to the Works. It is quite possible that a statutory undertaker may delay one site, affecting the access perhaps, while carrying out statutory obligations to lay pipe or cables to an adjacent site. That is not something for which the contractor would be entitled to an extension of time. This clause does not appear to apply to a statutory undertaker acting under its powers, if it is not acting as a duty. Neither does it apply to them when acting as directly employed contractors to the employer (in such a case, clause 2.26.5 would most likely apply).

Clause 2.26.7 Exceptionally adverse weather conditions

This event is worded so as to embrace unusually dry as well as unusually wet or frosty conditions. A long hot summer can cause great problems on site, particularly with such matters as the curing of concrete. The key word is 'exceptionally'. The weather must be exceptionally adverse in the light of the kind of weather usually encountered at that time of year or in that place. The contractor is expected to allow for the normal deviation in weather patterns. In order to decide whether the weather fits that description, meteorological reports are helpful. It is suggested that reports for the previous 10 years would be necessary to establish that the adversity was exceptional. In deciding whether a particular piece of exceptionally adverse weather is to be allowed as grounds for an extension of time, the employer must consider its effect on the works at the stage they have actually reached, not the stage they should have reached in accordance with some programme. Even if the weather is affecting the Works, because the contractor's own delay has prevented the building being sealed against the weather, the contractor is entitled to an extension of time if it appears likely the completion date will be exceeded as a result: *Walter Lawrence & Son Ltd* v. *Commercial Union Properties (UK) Ltd* (1984).

Clause 2.26.8 Loss or damage caused by specified perils

It is noteworthy that the contractor is not entitled to an extension of time for delay caused by all the insurance risks, but only the specified perils, which are noted as fire, lightning, explosion, storm, tempest, flood, escape of water from any water tank, apparatus or pipes, earthquake, aircraft and other aerial devices or articles dropped therefrom, riot and civil commotion, but excluding excepted risks, i.e. ionising radiations or contamination by radioactivity from any nuclear fuel or from any nuclear waste from the combustion of nuclear fuel, radioactive toxic explosive or other hazardous properties of any explosive nuclear assembly or nuclear component thereof, and pressure waves caused by aircraft or other aerial devices travelling at sonic or supersonic speeds.

The main difference between 'specified perils' and 'all risks' is the risks of impact, subsidence, theft or vandalism. Delay to the completion of the Works resulting from loss or damage caused by these risks must be dealt with by the contractor. This is a point which the contractor must take into account when required to obtain insurance cover under option A (see section 11.3).

Clause 2.26.9 Civil commotion or the use or threat of terrorism and the activity of authorities in dealing with it

An essential element of civil commotion is turbulence or tumult. It may amount to *force majeure*. It has been said to be more serious than a riot, but less than civil war: *Levy* v. *Assicurazione Generali* (1940). Unfortunately, terrorism needs no explanation. It seems that the threat must be real and not just a generalised concern.

Clause 2.26.10 Strike, lock-out affecting trades employed on the Works, preparing, manufacturing or transporting materials for the Works or persons designing the Works

Strikes by three kinds of persons are included: persons working on site, persons working off site getting things ready for site or delivering them, and persons designing the Works. It is uncommon for an independent consultant on the design team to have to deal with a strike. Strikes in the other two categories are more likely. The relevant events must be read strictly. The strike provision is broad enough to encompass both official and unofficial strikes, but it will not include a work to rule. A strike affecting goods required for the Works is included, but not a strike affecting deliveries of raw materials to a factory for the manufacture of goods required for the Works.

Clause 2.26.11 Government exercise of statutory powers

This event applies only where the UK government has exercised powers after the base date. The exercise must directly affect the Works by restricting the availability of labour, materials or fuel which are essential to the carrying out of the Works.

Clause 2.26.12 Delay in receiving necessary permission of a statutory body

This relevant event has already been considered in section 6.3. The contractor must have taken all practicable steps to reduce the delay. This event clearly refers to every kind of statutory permission or approval. Different buildings will attract differing regulations. Most buildings require planning permission and must satisfy the Building Regulations, but there are fire regulations, entertainment licences, water regulations and many more statutory controls which may apply. The contractor must be able to show that it has made any necessary applications at the appropriate time and that the reason for the delay is not something for which it is responsible, i.e. by not taking all practicable steps to reduce the delay. What are 'practicable steps' may appear to be self-evident. Useful guidance has been given in *Jordan* v. *Norfolk County Council* (1994) where the judge held that the term 'reasonably practicable' referred not just to physical practicability, but also to whether a course of action was practicable in the financial sense.

Clause 2.26.13 Force majeure

This is a term used in French law and it is broader than 'Act of God', referring to all circumstances independent of the will of man: *Lebeaupin* v. *R Crispin & Co* (1920). It is seldom called in aid by a contractor suffering delay, because most of the circumstances which obviously fall under this clause are already covered under other relevant events. Such events as civil commotion, government decrees, fire or exceptional weather conditions spring to mind.

In WCD 98, there used to be a relevant event dealing with the contractor's inability to secure labour and materials. It tended to be a source of difficulty in interpretation and it has been omitted from this contract with the result that the contractor is not entitled to any extension of time on these grounds whether or not the circumstances were foreseeable at the date the contract was executed.

8.4 Liquidated damages

8.4.1 General principles

It is open to the parties to a contract to agree upon a fixed sum of money which one will pay to the other in the case of a breach of contract. In the case of building contracts, it is usually stipulated that such a sum will be paid if the contractor fails to complete the Works by the date for completion in the contract particulars. This contract is no exception and the terms of such payment are to be found in clause 2.29.

The arrangement saves the parties the uncertainty and expense of a legal action to determine the damages payable for the breach. In the absence of clause 2.29, the employer would be left to recover whatever amount of unliquidated damages could be proven. The liquidated damages clause is often (incorrectly) referred to as a penalty clause. There is a significant difference between them. Liquidated damages must be a genuine pre-estimate of loss. That is to say that it must be a figure inserted into the contract by the employer to represent the best estimate that could be made, at the date the contract was executed, of the likely loss the employer would suffer if the completion was delayed. A penalty, however, is a punishment whose value bears no relation to the damages expected to be incurred. A penalty is not enforceable. It makes no difference what terminology is used, it is the reality which is important. Certain guidelines have been set out for the recognition of a penalty, notably in *Dunlop Pneumatic Tyre Co Ltd* v. *New Garage & Motor Co Ltd* (1915) per Lord Dunedin:

> 'It will be held to be a penalty if the sum stipulated for is extravagant and unconscionable in amount in comparison with the greatest loss which could conceivably be proved to have followed from the breach . . . It will be held to be a penalty if the breach consists only in not paying a sum of money, and the sum stipulated is a sum greater than the sum which ought to have been paid . . . There is a presumption (but no more) that it is a penalty when a single lump sum is made payable by way of compensation, on the occurrence of one or more or all of several events, some of which may occasion serious and others but trifling damages . . . On the other hand . . . it is no obstacle to the sum stipulated being a genuine pre-estimate of damage that the consequences of the breach are such as to make precise pre-estimation almost an impossibility. On the contrary, that is just the situation when it is probable that pre-estimated damage was the true bargain between the parties.'

In practice, many liquidated damages provisions are so badly expressed that they are either inconsistent with other clauses in the contract: *Bramall and Ogden Ltd* v. *Sheffield City Council* (1985), or in operation they become penalties: *Stanor Electric Ltd* v. *R Mansell Ltd* (1988). If such a clause is held to be a penalty, all is not lost so far as the employer is concerned who is left with the common law remedy of suing for such damages as can be proven. Although the point has not been conclusively settled, it appears that the employer would be unable to recover more than the amount set down as a penalty. Any other conclusion would be inequitable.

Liquidated damages are recoverable without proof of loss. It matters not that the employer has lost less than expected, lost nothing at all, or even that the employer has made a profit as a result of the contractor's late completion. The employer is entitled to the liquidated damages in each of these instances (*BFI Group of Companies Ltd* v. *DCB Integration Systems Ltd* (1987)) even where the employer is in occupation of the building before practical completion (*Impresa Casteli SpA* v. *Cola Holdings Ltd* (2002)). The amount of liquidated damages is to be inserted in the contract particulars as £ . . . per. . . . If the space is left blank, the law does not permit the employer to produce evidence regarding what was intended to be included: *Kemp* v. *Rose* (1858). In such circumstances, the employer would not be entitled to recover any liquidated damages, but it is probable that actual loss could be recovered on the same basis as if the sum indicated was a penalty, except that there would be no ceiling on the possible recovery. If, however, 'Nil' is inserted, that figure would signify the amount of damages recoverable per day or per week and the employer would be unable to sue for damages at common law, because the provision for liquidated damages is exhaustive of the employer's rights to damages: *Temloc Ltd* v. *Errill Properties Ltd* (1987). The courts increasingly take a pragmatic view of liquidated damages clauses. In *Alfred McAlpine Capital Projects Ltd* v. *Tilebox Ltd* (2005), when upholding a liquidated damages clause the judge said:

> 'This court, following the lead set by higher courts, is predisposed where possible to uphold contractual terms which fix the level of damages. This predisposition is somewhat stronger in the present case for the following reason: the building contract . . . is a commercial contract made between two parties of comparable bargaining power.'

The courts are willing to accept a liquidated damages provision which is graduated to reflect the seriousness of the particular breach: *North Sea Ventilation Ltd* v. *Consafe Engineering (UK) Ltd* (2004).

8.4.2 Contract provisions

Clause 2.28 provides that if the contractor fails to complete the Works by the completion date or by any extended date, the employer must issue a written notice to the contractor. The notice is important. It is a pre-condition to the right of the employer to recover liquidated damages. It does not have the same weight as an architect's certificate under clause 2.31 of SBC and the employer cannot rely upon it so as to deduct liquidated damages from retention money so as to extinguish the fund: *J F Finnegan Ltd* v. *Ford Seller Morris Developments Ltd (No. 1)* (1991). If the employer makes a further extension of time after issuing the notice, clause 2.28 makes clear that the original notice is cancelled and a new one must be issued. This follows the judgment in *A Bell & Son (Paddington) Ltd* v. *CBF Residential Care & Housing Association* (1989).

The contractor must pay or allow the liquidated damages and the employer may either deduct them from any monies due or to become due to the contractor, or they may recovered as a debt. Not surprisingly, liquidated damages are invariably deducted. Following the provisions in the Housing Grants, Construction and Regeneration Act 1996 and in what seems to be an excess of caution, the recovery process has been made quite complicated. There are three pre-conditions which must be satisfied before recovery of liquidated damages can take place:

- The employer must have issued a non-completion notice under clause 2.28; and
- The employer must issue a written notice to the contractor, informing it that the employer may require payment of, or may withhold or deduct, the liquidated damages (clause 2.29.1.2), which will be referred to below for clarity as the 'second notice'.

- The employer must issue a written notice or requirement requiring payment or notifying deduction (clause 2.29.2), which will be referred to below for clarity as the 'third notice'.

The second notice may be served at any time after the non-completion notice, but not later than the date on which the final account and final statement become conclusive. In a similar way, the third notice may be served at any time after the second notice, but not later than 5 days before the final date for payment under clause 4.12. This third notice, if notifying deduction, can take the place of the withholding notices which must be served under clauses 4.10.4 or 4.12.9, provided that it satisfies the criteria for withholding notices set out in those clauses. Otherwise, a separate withholding notice must be served before liquidated damages can be deducted.

Clause 2.29.4 provides that the employer need only serve the second notice once. It remains effective, despite the issue of further non-completion certificates, unless the employer withdraws it. Since the decision to deduct liquidated damages rests with the employer, it is unlikely that, in practice, the second notice would ever be withdrawn. If the employer decided not to deduct damages, the matter would simply be allowed to rest.

Some doubt has been thrown on the precise form to be taken by the employer's written requirement for payment. In the *Bell* case, considering a similar clause under the JCT 80 Standard Form, Judge John Newey QC stated:

> 'There can be no doubt that a certificate of failure to complete given under clause 24.1 and a written requirement of payment or allowance under the middle part of clause 24.2.1 were conditions precedent to the making of deductions on account of liquidated damages or recovery of them under the latter part of clause 24.2.1.'

This seems perfectly clear, but in *Jarvis Brent Ltd* v. *Rowlinson Construction Ltd* (1990), again considering JCT 80, it was held that the written requirement was satisfied by a letter, written by the quantity surveyor and forwarded to the contractor, which stated the amount which the employer was entitled to deduct; alternatively, it was stated that the cheques issued by the employer from which liquidated damages had been deducted constituted such written requirements. The judge went on to consider whether the written requirement was indeed a condition precedent and came to the conclusion that it was not. All that was necessary was that the contractor should be in no doubt that the employer intended to make the deduction. In *Holloway Holdings Ltd* v. *Archway Business Centre Ltd* (1992) a similar clause in IFC 84 was considered and it was again held:

> 'For (the employer) to be able to deduct liquidated damages there must both be a certificate from the Architect and a written request to (the contractor) from (the employer).'

The matter was finally clarified by a decision of the Court of Appeal in *J J Finnegan Ltd* v. *Community Housing Association Ltd* (1995) where the court held that the decision in *Bell* was correct and that the employer's written requirement was a condition precedent to the deduction of liquidated damages. Only two things must be specified in the requirement and they are:

- Whether the employer is claiming a payment or a deduction of the liquidated damages; and
- Whether the requirement relates to the whole or part of the total liquidated damages.

Clause 2.29.1.2 of DB is, of course, in very similar terms. The current clause 2.29.1.2 places a duty on the employer to require the payment or allowance in writing before making a deduction of liquidated damages. Quite apart from the plain words of the contract (which also occur in SBC, IC and ICD), the new clause 2.29.4 emphasises that a requirement which has been stated in writing remains effective even if the employer issues further non-completion notices. It is difficult to understand why it is thought necessary for three notices to be given, and the drafting could be clearer to say the least. However, whatever the intention may have been, it appears that, on the present wording, a minimum of three notices is now essential before recovery of liquidated damages under the contract is permitted.

The amount which the employer may deduct is to be calculated by reference to the rate stated in the contract particulars. The employer is free to reduce the rate, but not to increase it. That is expressly stated in clauses 2.29.2.1 and 2.29.2.2: 'or at such lesser rate . . .'. Clause 2.29.1 now makes clear that the employer need not wait until practical completion before deducting liquidated damages. The employer may start to deduct them as soon as the relevant notices have been issued. In practice, such deductions usually commence from the first payment thereafter. Clause 2.29.3 clarifies that if a later completion date is fixed, the employer must repay the relevant liquidated damages. There is no provision for payment of interest on such repaid liquidated damages. The contractor's claim for such interest would fail unless it could be shown that the deduction of liquidated damages was itself a breach of contract (see section 5.4.6).

In section 7.3, the complication introduced into practical completion by the requirement to comply with clauses 2.37 and 3.18.4 was discussed. Although undoubtedly the parties are agreeing to take as practical completion for all the purposes of the contract (including the recovery of liquidated damages) the date specified in clause 2.27, it is possible that practical completion of the physical Works may take place days or even weeks before compliance with clauses 2.37 and 3.18.4 is finally achieved. Since compliance may be complete except for some relatively minor parts of the health and safety file, it is arguable that where physical practical completion has been achieved as a matter of fact and where there are minor parts of the health and safety file outstanding, the liquidated damages amount may be a penalty.

Chapter 9
Financial Claims

9.1 Types of claim

The dictionary definition of 'claim' is 'an assertion of a right'. All too often, claims in the building industry are looked upon, and sometimes are, assertions of presumed rights which have no foundation in reality. In the context of building, a claim is usually a claim for money outside the contractual machinery for valuing the work. It may also be a claim for extension of time. There are three kinds of money claim commonly made by contractors:

- *Contractual claims:* These are claims which are made under the express provisions of the contract. They are outside the normal contractual mechanism for valuation of work carried out, and DB deals with them in clauses 3.17, 4.19, 4.20 and supplemental provision para. 5. These clauses set out specific grounds on which the contractor can claim and they also specify the precise manner in which it must proceed with its claim. Many of the grounds would not allow the contractor to claim extra payment if they were not in the contract, because they are not breaches. In general, where the contract specifies certain requirements which must be satisfied in order to entitle the contractor to recover direct loss and/or expense, failure to satisfy those requirements will prevent recovery. These are the only kinds of claims which the employer's agent is empowered to deal with and the only kinds of claims which the employer may deal with under the contract.
- *Common law claims:* These claims arise outside the express provisions of the contract. They may be claims in tort or for breach of the contract's express or implied terms. They may be claims that there is no concluded contract and that the contractor is, therefore, entitled to be paid on a *quantum meruit* basis. Nothing in DB prevents the contractor proceeding with any such claims through the dispute resolution procedure specified in the contract, provided the action is brought before the final account and final statement become conclusive. Such claims are sometimes termed ex- or extra-contractual claims. The employer's agent has no power to deal with them. The employer may deal with them, of course, but not under the contractual claims procedures unless the contractor agrees.
- Ex gratia *claims:* These are sometimes known as 'hardship' claims and they have no legal foundation. The contractor has no entitlement as a right. A contractor may advance this sort of claim as a last resort when it knows that it has suffered a large loss without there being any fault on the part of the employer. It is entirely up to the employer whether to meet this claim in full or at all and it is suggested that such claims will not normally be successful unless there are special circumstances which make it advantageous for the employer to take this step.

Contractors who make claims are often unfairly dubbed 'claims conscious'. A contractor who makes justified claims is simply efficient and takes advantage of the procedures which are in the contract or in the common law precisely for the purpose of providing it with a remedy under the particular circumstances when it is right that it should have one. The contractors of which one should beware are those who make inflated or spurious claims as part of their normal approach to any project.

9.2 Application for direct loss and/or expense

DB sets out a procedure which the contractor must observe if it wishes to recover direct loss and/or expense under the terms of the contract. The contractor has no duty to carry out the procedure, but if it does not do so, it cannot recover loss and/or expense. If it wishes to set the procedure in motion, the contractor must make application under clause 4.19. The application must be in writing and it must state that the contractor has incurred or is likely to incur direct loss and/or expense for which it would not be reimbursed under any other provision of the contract. That is important. If the contractor is entitled to recover under some other provision, for example clause 5, it cannot recover under this clause.

The contractor must make the application as soon as it is apparent or should reasonably have become apparent that the regular progress of the Works has been or was likely to be affected. The contractor must act promptly. Although it may not be able to give details, it will very quickly know whether regular progress is likely to be affected by any particular circumstance. If it fails to act promptly, it will lose its entitlement to reimbursement of loss and/or expense under the contract.

The grounds upon which the contractor founds its application must be stated. They may be deferment of giving possession of the site (where clause 2.4 is stated in the contract particulars to apply) or regular progress being materially affected by one of the relevant matters listed in clause 4.20. Use of the phrase 'materially affected' makes clear that the contractor may only apply when the affect on progress is substantial. Trivial disturbances or delays must be absorbed. The degree of 'affectation' which can be categorised as 'material' may be disputed.

It should be noted that, in this contract, the criterion is that regular progress has been *or* is likely to be materially affected. In clause 26.1 of WCD 98, the two parts of the criterion were linked by 'and'. On a strict reading of the 1998 clause it appeared that the contractor could not make application unless it made it at a time when regular progress not only had been (i.e. in the past) materially affected but also would be affected in the future. In DB, the 'and' is changed to 'or' so that a contractor can choose whether to apply after the material affectation or before it. However, that apparent freedom is modified by clause 4.19.1 which stipulates that the application must be made as soon as it has become, or ought reasonably to have become, apparent that regular progress is likely to be affected. Applications made 6 months after practical completion would not comply with this requirement.

Clause 4.19.2 provides that the contractor must support its application and the amount of loss and/or expense by giving the employer the information and details which the employer may reasonably require. It is to be noted that it is the employer's requirement, not the details, which are to be reasonable. The contractor is not obliged to provide supporting information unless this is specifically required by the employer. In *London Borough of Merton* v. *Stanley Hugh Leach Ltd* (1985), Mr Justice Vinelott famously said about the contractor applying for loss and/or expense:

> 'He must make his application within a reasonable time: it must not be made so late that, for instance, the architect can no longer form a competent opinion on the matters on which he is required to form an opinion or satisfy himself that the contractor has suffered the loss or expense claimed. But in considering whether the contractor has acted reasonably and with reasonable expedition it must be borne in mind that the architect is not a stranger to the work and may in some cases have a very detailed knowledge of the progress of the work and of the contractor's planning.'

The contract under consideration was JCT 63 with provision for an architect, and the loss and/or expense clause was somewhat different in wording, but if the employer has engaged an agent, it is reasonable to suppose that the agent will not be a 'stranger to the work' either. Whether the employer's request for any particular piece of information is reasonable should be considered in that light.

There is another difference between the subject of Mr Justice Vinelott's comments and this form. Under JCT 63, it is the responsibility of the architect to decide whether an application is valid and then to ascertain, or to instruct the quantity surveyor to ascertain, the loss and/or expense. In DB, it is simply stated that the amount of the loss and/or expense incurred or being incurred by the contractor is to be added to the contract sum. Because it is the contractor who is to make application for interim payment under clause 4.9, the contractor must also calculate the amount of loss and/or expense it is suffering. The employer's right to 'reasonably require' information is for the purpose of checking the contractor's application, not to ascertain it in the first instance. The employer's agent is not charged with holding the balance between the parties, neither it seems does the agent owe the employer a duty to act fairly (see section 5.1). In this context, the ascertainment of loss and/or expense under DB has a changed emphasis. Whereas under SBC and most other standard forms in the traditional mould, it is the architect who determines the amount payable to the contractor as loss and/or expense, and it is for the contractor to successfully dispute the amount if it can, under this form the burden of showing that the contractor's ascertainment is wrong is laid on the employer. This gives the contractor a distinct advantage.

What the contractor can recover as direct loss and/or expense is established as the same as could be recovered as damages at common law under the rules set out in *Hadley* v. *Baxendale* (1854). That is the loss which the parties could reasonably foresee would be the direct result of the breach of contract, considered at the date when the contract was entered into. There are no particular limits to the losses which the contractor can recover as direct loss and/or expense, provided that they are within the reasonable contemplation of the parties and flow directly from the event relied on. What the contractor can include will depend on the facts in each case.

Common heads of claim include the following:

- Plant and labour inefficiency
- Increases in cost
- Increases in head office overheads
- Acceleration

- Establishment costs during any period of delay
- Loss of profit (if the contractor can show that it could earn the profit elsewhere)
- Interest and financing charges: *F G Minter* v. *Welsh Health Technical Services Organisation* (1980).

The contractor, however, must be able to show how it has incurred the loss and/or expense in some detail. It is not sufficient to list contractual grounds in a general way ('We have been delayed and disrupted due to extra work, lack of approvals and interference by other contractors'). The disruptive matter must be specified and the effect noted in each case: *Wharf Properties Ltd* v. *Eric Cumine Associates* (1991). Where the consequences of the various matters interact in a complex way, it may be difficult or impossible to separate the evaluation. In that case, the contractor is entitled to put forward a composite calculation: *J Crosby & Sons Ltd* v. *Portland Urban District Council* (1967). There is considerable misunderstanding of this point in the industry. In simple terms, the causes and effects must be individually identified, but in certain circumstances the calculation of resultant loss and/or expense may be carried out on a global basis: see *Imperial Chemical Industries plc* v. *Bovis Construction Ltd* (1992).

In that case, ICI brought proceedings against Bovis, the management contractor, as well as against the architects and consulting engineers. ICI's statement of claim contained several pages of allegations against each of the defendants, but these were not otherwise particularised and no attempt was made to link any alleged breach to a particular loss. In paragraph 21 a global claim was made against the defendants for a sum of £19 million and professional fees.

The pleadings were amended but still not, it seemed – at least to the defendants – adequately particularised. Judge Fox-Andrews QC ordered ICI to serve a Scott Schedule setting out the alleged complaint, against whom it was made, which clause of the contract had been breached and the alleged factual consequences of each breach. They complied with his order but still not to the satisfaction of the defendants; many of the items were pleaded on a global basis.

This practice first received the approval of the Courts in *Crosby* v. *Portland* and was blessed by Mr Justice Vinelott in *Merton* v. *Leach*. The global approach is only acceptable, it has been said, where a claim depends on 'an extremely complex interaction in the consequences of various denials, suspensions and variations, [where] it may be difficult or even impossible to make an accurate apportionment of the total extra cost between several causative elements'.

Under the 'global approach' there must (self-evidently) be no duplication, and as was noted in *Leach* a global award

> 'can only be made in a case where the loss or expense attributable to each head of claim cannot in reality be separated and . . . can only be made where apart from that practical impossibility the conditions which have to be satisfied before an award can be made have been satisfied in relation to each head of claim'.

It is not an excuse for sloppy pleading or for failure to prove one's case, as was emphasised by the Privy Council in *Wharf* v. *Cumine*, although their Lordships expressed no reservations about the correctness of the global approach. What the Privy Council actually said was this:

> 'What those cases actually establish is no more than this, that in cases where the full extent of extra costs incurred through delay depend upon a complex interaction between the consequence of various events, so that it may be difficult to make an accurate apportionment of the total extra costs, it may be proper for the arbitrator to make individual financial awards in respect of claims which can conveniently be dealt with in isolation and a supplementary award in respect of the financial consequences of the remainder as a composite whole. *This, however, has no bearing upon the obligation of a plaintiff to plead his case with such particularity as is sufficient to alert the opposing party of the case which is going to be made against him at the trial.*' [my emphasis]

Having surveyed these cases, Judge Fox-Andrews QC considered the Scott Schedule in detail. He found some of the items 'objectionable' and another 'hopelessly inadequate'.

> 'In respect of the many hundreds of items itemised on 101 pages, the financial consequences of which are always stated to be the same, namely "The total cost of abortive work amounts to £840,211 as particularised in Appendix 7 . . ." ICI's case was that the various events set out all contributed to the sums claimed, with no actual apportionment being possible . . . I find that it is palpable nonsense that £840,211 would be the cost of repositioning a bell. It is important to appreciate that whilst a pleading may take a particular form where a number of interactive events give rise to delay and disruption, the same does not appear to me to apply to many of the items [listed by ICI]. *The financial consequences of each breach, where possible, must be pleaded and the necessary nexus shown.*' [my emphasis]

In the event, the judge decided that since a great deal of work had been done by ICI and their advisors, they should not be debarred from pursuing their claim. However, he ordered that a fresh Scott Schedule giving the necessary particulars and showing the causal nexus should be served. A totally new and revised document was required.

If it is possible to evaluate the delaying or disruptive effects of individual causes, this must be done, leaving only the balance of the delays for which the employer is alleged to be responsible to be swept up by the 'global approach'. It is clear that the contractor may put its claim in whatever form it wishes, but claims on a global basis may suffer severe evidential problems: *GMTC Tools & Equipment* v. *Yuasa Warwick Machinery* (1994). The judgment in *How Engineering Services* v. *Lindner Ceilings Partitions* (1995) is very instructive on this point. There, in a careful judgment which is of general application, the court said that the claim must be intelligible and it must identify the loss, the reason for it and why the other party has an enforceable obligation to compensate for the loss. The claim should tie breaches to contract terms which should identified. Cause and effect should be linked. Although there is no obligation on the contractor to break down the loss to identify the sum claimed for each specific breach, failure to do so will create an 'all or nothing' claim which will completely fail if some of the events cited are not substantiated. The court concluded by stating that a global claim must identify two things. The first was a means by which the loss is to be calculated if some of the events are not established. There should be some kind of realistic formula to achieve the scaling down of the claim. The second was a means of scaling down the claim to take account of the various other factors such as defects, inefficiencies or events which are at the contractor's risk. The calculation of loss should be carried out accordingly.

In clause 4.19, the words 'which has been or *is being* incurred' [my emphasis] indicate that the contractor need not wait until it has finished incurring loss and/or expense before the amount 'shall be added to the Contract Sum'. The matter is put beyond doubt by clause 4.21 which states that amounts ascertained 'from time to time' must be added to the contract sum. Clause 4.3 states that as soon as any amount is ascertained *in whole or in part* it must be included in the next interim payment. The scheme of clause 4.19 is straightforward. Briefly, it is for the contractor to apply as soon as it realises that it is suffering loss and expense. It must include the appropriate sums, as they can be ascertained, in applications for payment and if the employer requires supporting information, it must be provided. The contractor is entitled to have the sums included in the next payment after ascertainment so that it is not unreasonably kept out of its money.

Clause 4.22 states plainly that the provisions of clauses 4.19–4.21 are without prejudice to other rights and remedies which the contractor may possess. This means that the contractor may choose to exercise its common law rights instead of relying on clause 4.19, or it may apply under clause 4.19 in order to recover whatever it can and top up this amount later with a common law claim. It also means that the contractor is not tied to the times and procedures set out in clause 4.19 except in one important respect. Most claims at common law will be based on the employer's alleged breaches of express or implied terms of the contract. Several of the grounds set out in clause 4.20 are not breaches of contract. For example, it cannot be a breach of contract for the employer to issue an instruction requiring a change or opening up of the Works, because the contract gives the employer power to issue just such an instruction. Therefore, if the contractor does not satisfy the provisions of clause 4.19 in respect of grounds like these, which are not breaches of contract, it will be unable to bring a successful action for damages on these grounds at common law.

A contractor who intends to bring a common law claim should also note the provisions of clause 1.9.1.3 (see Chapter 10, Section 10.6). When the final statement has become conclusive under the terms of the contract, it is conclusive evidence, among other things, that where the contractor has received payment under the provisions of clause 4.19, in broad terms it finally settles any claims the contractor has or may have in the future. There are some significant points to note about clause 1.9.1.3, which contains this provision:

- The finality is expressed as referring to claims for breach of contract, duty of care (tort), statutory duty or otherwise. It is suggested that 'otherwise' would be interpreted *ejusdem generis* in this instance to refer to claims of the same class as those expressly mentioned.
- The finality refers only to claims arising out of the occurrence of any relevant matters in clause 4.20. Therefore, it is open to a contractor to bring a subsequent common law claim concerning some breach etc. which is not included in clause 4.20. The opportunities for such claims are apparently limitless. But it should be noted that breach of most of the employer's obligations under the contract are included in clause 4.20.5.
- The inclusion of the words 'if any' after reference to the reimbursement of direct loss and/or expense, may restrict the operation of this clause to those circumstances where there has been some payment under clause 4.19, although the draftsperson probably wanted the contrary effect. It may be that if there has been

no payment under clause 4.19, the contractor is free to pursue its common law claims after the date at which the final statement becomes conclusive. A contractor wishing to bring a common law claim must be careful, but clause 1.9.1.3 may not be so final as is generally thought.

9.3 Supplemental provision procedures (para. 5)

These procedures greatly resemble the procedures in the Association of Consultant Architects Form of Building Agreement (ACA 3). They are very straightforward and sensible and promise advantages to both parties if properly operated. The first thing to note is that clauses 4.19–4.22 are modified, but not superseded, by this provision. The second thing is that loss and/or expense which is dealt with under para. 4 (valuation of change instructions) is excluded from treatment under this clause. That is simply to avoid any possibility of the contractor obtaining double recovery.

The procedure is triggered as soon as the contractor is entitled to have some direct loss and/or expense added to the contract sum. The contractor must include an estimate of the amount in its next application for payment. The amount must refer to the period immediately before the application. Therefore, if the contractor is receiving payments at monthly intervals, it must include in its estimate the whole of the amount it requires to represent its direct loss and/or expense during that month preceding payment. This places an obligation on the contractor to act swiftly when it becomes aware that it is incurring losses. Besides making the normal application in accordance with clause 4.19, it must calculate the loss and/or expense and insert it into the next payment application (para. 5.2). It is termed an 'estimate' rather than an ascertainment, because in many instances, the contractor will not be able to calculate a precise figure.

In some cases, the contractor will incur the loss and/or expense over a long period. Para. 5.3 stipulates that the contractor must continue to submit estimates for as long as necessary, each estimate referable to the preceding period. Therefore, in the example noted above, the contractor would submit estimates every month until the loss ended and each estimate would refer to the preceding month (para. 5.3).

The contractor's estimates are dealt with in accordance with para. 5.4. The employer has 21 days from receipt of the contractor's estimate in which to give a written notice to the contractor. The employer may request information reasonably required to support the contractor's estimate, but the employer must request and receive such information within the 21 days. The employer may not delay giving notice on the grounds that information has not been received. Of course, the content of the employer's notice will doubtless depend very much on the information received. The employer may not simply reject the contractor's estimate, but may state one of the following options in the notice.

The employer may:

- Accept the estimate.
- State a wish to negotiate the amount. No time limit is given for the negotiations and it is suggested that it is in the interests of the parties to insert a short time limit; 7 days is not unreasonable. If agreement cannot be reached, the provisions

of clause 4.19 apply and the employer must refer the issue for a decision by the adjudicator.
- Simply state that the provisions of clause 4.19 apply.

The contract is silent as to what follows if the issue is referred to adjudication. However, if agreement is reached on the amount of the addition to the contract sum, the sum must be added and no further sums may be added for that particular matter in clauses 4.19 and 4.20 and in that period. There is an element of rough justice here in that both contractor and employer are held by tight time restraints and the contractor's finally agreed estimates may be somewhat wide of the mark. The idea is clearly that the contractor receives payment as soon as practicable after the event and in return the employer has the certainty that there will be no more claims for the effects of any clause 4.20 relevant matter during that period.

How does it work in practice? Assuming that the contractor submits its applications for payment every 4 weeks, the contractor will submit its estimate with its general application for payment under clause 4.9. The employer has only 14 days in which to make payment following an application (clause 4.10.1), so the contractor will not receive payment then, because the employer has 21 days to issue the notice. If the employer issues the notice on the last of the 21 days, if the employer opts to negotiate and if there is a time limit of 7 days imposed, agreement should be reached (or not reached) by the date of the next application, and payment of the estimate can be made 14 days thereafter. If the employer opts to revert to clause 4.19, the contractor is not disadvantaged because it is simply in the same position as it would be if para. 5 were not in the contract so far as that particular claim is concerned. The contractor is not then precluded from taking a reasonable time to calculate its precise loss and/or expense and to add the extra amount to its next application. As can be seen from clause 4.19, the employer is not in the position of ascertaining the amount due to the contractor, but an employer who disputes it must prove on the balance of probabilities that the contractor's ascertainment is wrong.

There is a sting in the tail (para. 5.6). If the contractor fails to submit its estimates as required under para. 5.2 and para. 5.3, para. 5 ceases to apply. The contractor's direct loss and/or expense is to be dealt with under clause 4.19, but the amounts are not payable until the final account and final statement are agreed. Under these circumstances, the contractor is not entitled to any interest or financing charges incurred before the issue of the final account or final statement. It is, therefore, important for the contractor to rigidly observe the rules laid down in this clause if it wishes to secure maximum advantage. Theoretically, the contractor should have few claims under this form of contract, because it has complete control over the design and construction. Unfortunately, experience shows that employers cannot resist changes during the progress of the Works, as well as the appointment of direct contractors to undertake special work. The result will only be justified loss and/or expense claims from the contractor.

9.4 *Grounds for direct loss and/or expense*

The grounds on which the contractor may make application for direct loss and/or expense are contained in clause 4.19 and 4.20. One is included in clause 4.19 and the remainder are referred to as 'Relevant Matters' and are contained in clause 4.20. In many instances, they echo relevant events in clause 2.26. This has doubtless led to the erroneous view that it is necessary for the contractor to obtain an extension of time under the appropriate clause before it is entitled to apply for loss and/or expense resulting from the same occurrence. That this approach is wrong is clear from the judgment in *H Fairweather & Co Ltd* v. *London Borough of Wandsworth* (1987), when this question was considered in relation to JCT 63 and Judge Fox-Andrews QC said:

> 'But I do not consider that the obtaining of an extension of time under (the extension of time clause) is a condition precedent to recovering loss and expense under (the loss and expense clause).'

The judgment in *Methodist Homes Housing Association Ltd* v. *Messrs Scott & McIntosh* (1997) is to similar effect. The grounds are as follows.

Clause 4.19 Deferment of possession

Clause 2.4 must be stated in the contract particulars to apply or the employer has no power to defer possession. If it is not stated to apply, but the contractor has not received possession on the due date, the employer is in breach of contract and the contractor is entitled to damages at common law and probably to treat the contract as repudiated under certain circumstances if the delay continues for a substantial period. However, the contract now provides a remedy under the relevant matter in clause 4.20.5. This is the sensible option if the delay to possession exceeds what is permitted under clause 2.4 but is not outrageously long. Deferment of possession will obviously cause delay. To what extent it also causes the contractor to incur direct loss and/or expense is for the contractor to demonstrate.

Clause 4.20.1 Changes

Any instruction under clause 3.9 may have financial repercussions beyond the ones provided for in the valuation rules. When that is the case, the contractor can claim under this head. This is also the head under which the contractor may be able to make application following a clause 2.15.2.1 situation where a change in statutory requirements after base date is to be treated as an instruction of the employer requiring a change. Where supplemental provision para. 4 applies, all the effects of the instruction are to be included in the contractor's estimate.

Clause 4.20.2 Instructions of the employer

Under clause 3.10, the employer may issue instructions to postpone any design or construction work (see section 5.3). It has been held that under the provisions of JCT 63 the architect may issue instructions which are effectively postponement instructions, although they were not issued expressly pursuant to the postponement clause: *Holland Hannen & Cubitts (Northern) Ltd* v. *Welsh Health Technical Services Organisation and Another* (1981). This possibility must not be discounted under this form of contract (substituting employer for architect), but unlike JCT 63, the contractor is unable to recover under these grounds for such instructions. Only employer's instructions issued 'under clause 3.10' can be considered. For the equivalent of the *Holland* situation, the contractor might argue that all postponement instructions must be pursuant to clause 3.10 or they are not empowered under the contract and the contractor cannot comply.

Potentially the most significant employer's instruction is regarding expenditure of a provisional sum. Such sums must be included in the Employer's Requirements, but they are often given with little explanation. The contractor has an obligation to use reasonable care in programming its work so as to achieve completion by the due date, and it may be thought that it must take the work in provisional sums into account so far as it is able. Realistically, the contractor will seldom be able to judge the work in a provisional sum with any accuracy and is entitled to make little or no allowance in those circumstances. Indeed, it is suggested that under this form of contract the contractor is under no duty to take the work in provisional sums into account at all. The result is probably that an instruction to expend a provisional sum will have effect as a simple addition of work and/or materials.

The employer's power to order opening up and/or testing is contained in clause 3.12 (see section 5.3.2). In order to qualify as grounds for recovery of direct loss and/or expense, the work, materials or goods must be shown to be in accordance with the contract. In such cases it is almost inevitable that the contractor will have a claim. In typical circumstances, there will be an instruction under clause 3.12 stipulating that the employer's agent must be present at the opening up. When the contractor opens up there will be a pause to allow the parties to decide whether the work conforms to the contract. It may take some time before this stage is completed, especially if tests have to be taken to clarify the point. If it is decided that the work does conform, the disturbed work must be made good and only then can any work in that area proceed. It is clear that the decision to order opening up under clause 3.12 must be taken with a degree of circumspection related to the importance of the possible defect, its position in the Works and the effect of the instruction on progress.

Clause 4.20.3 Suspension of contractor's obligations

This matter refers to the same grounds as the equivalent clause 2.26.4, but the wording is not identical. It should be noted that in order to recover loss and/or expense, the suspension must not be frivolous or vexatious. That immediately begs the question whether an extension of time may be made if the suspension is frivolous or vexatious. Common sense suggests the contrary. Suspension carried out frivolously or vexatiously would be invalid. Clause 2.26.4 is merely putting into effect a legislative provision. There is nothing in the legislation to suggest that the contractor may suspend frivolously or vexatiously and the same can be said of clause 4.11. The introduction of this quite unnecessary proviso can only serve on occasion to provoke dispute.

Clause 4.20.4 Delay in receiving development control permission

The permission must be necessary for the Works to be carried out or to proceed and the contractor must have taken all practicable steps to avoid or reduce the delay. This probably means no more than that the contractor must make whatever application is necessary for permission as soon as it is appropriate to do so. Development control requirements have been discussed in section 6.3. Although broader in meaning than merely planning requirements, they are certainly not as wide in meaning as statutory requirements, of which they form part. Delays attributable to statutory requirements in general may perhaps be caught by clause 4.20.1, above, if they relate to a change in the requirements after base date, as described in clause 2.15.2.1. In most instances, claims will be made under this head when the obtaining of planning permission and the like has been left to the contractor. The contractor cannot seek permission until after it has secured the contract and at that stage it is difficult to avoid delay. It is no doubt possible to delete the contractor's entitlement as a result of delay in obtaining permission, but it should be noted that several clauses interrelate on this topic and there is great danger of failing to completely deal with the matter: *Update Construction Pty Ltd* v. *Rozelle Child Care Centre Ltd* (1992).

Clause 4.20.5 Employer impediment, prevention or default

This clause is identical to clause 2.26.5 and the comments in section 8.3 are relevant.

9.5 Antiquities

Clauses 3.15–3.17 deal with the situation if the contractor finds any fossils or other objects of interest or value, and provides that the employer may give instructions regarding the examination, excavation or removal of the object. In any event, the contractor must use its best endeavours not to disturb the object and to preserve it in its exact position even if this involves cessation of work in whole or in part. Clause 3.17 provides that if the contractor suffers loss and/or expense as a result of complying with the employer's instructions, it is to be added to the contract sum. The only stipulation is that the contractor would not be reimbursed under any other clause. There is no requirement that the contractor must make application or supply information, and recovery under this clause is not linked to clause 4.19. Indeed, it appears deliberately not to have been so linked, because it would have been a relatively simple task to include antiquities as one of the relevant matters under clause 4.20. The contractor merely has to calculate its loss and/or expense and to submit it with its application for payment under clause 4.9. If supplemental provision para. 5 applies, the contractor must include these amounts in its claims under para. 5.

Chapter 10
Payment

10.1 Contract sum

DB is a lump sum contract; that is to say that the contractor carries out completion of the design and the whole of the construction for a stated amount of money which is to be paid by the employer. The fact that there are clauses which allow the amount to be varied is irrelevant. The important point is that the original sum is specified as being for a given amount of work. This is to be contrasted with a contract which expressly provides for re-measurement.

Clause 1.1 defines the 'contract sum' as 'the sum stated in article 2'. Article 2 contains a blank space in which the parties are to insert the figure upon which they agree. Clause 4.1 importantly stipulates that the contract sum must not be adjusted or altered in any way except in accordance with the express provisions of the contract. So, for example, a contractor who is having financial difficulties cannot call in aid an implied term allowing any such adjustment. The particular provisions of the contract which do allow such adjustment are noted in Fig. 10.1. Clause 4.3 provides that where it is stipulated that any amount is to be added to or deducted from the contract sum, or if an adjustment is to be made to the sum, as soon as ascertainment of the whole or part of the amount has taken place, it is to be included in the next payment. This clause is virtually identical to the provision in SBC, but it assumes particular importance in the absence of valuation by a quantity surveyor.

In order to understand the way in which the contract works, it is important to remember that the value of the contract sum never changes. It may be adjusted, but it then becomes the contract sum which has been adjusted and in no sense is it a new contract sum. Article 2 is very clear about this. A sum is to be inserted and that sum is referred to as 'the Contract Sum'. This is stated to be the sum which the employer will pay to the contractor or such other sum as shall become payable under the contract. Once the contract sum is adjusted it becomes 'such other sum'. This appears to be an obvious point, but many construction professionals are becoming confused by the concept and fail to appreciate the importance of having an immutable figure as the contract sum.

Provisions for dealing with inconsistencies within the Employer's Requirements and the Contractor's Proposals are given in clause 2.14, but there are no provisions for dealing with errors in the contract sum or within the Contract Sum Analysis. Where the employer has, unusually, included bills of quantities in the Employer's Requirements and supplemental provision para. 3 applies, any errors in description or quantity must be corrected by the employer and treated as if they were changes in the Employer's Requirements. If the contractor makes an error in tendering by overlooking items or by making a mistake in adding up totals, it must

Clause	Adjustment
2.5.2	Additional premium for early use by employer.
2.10.1	Divergence between Employer's Requirements and the definition of the site boundary.
2.12.2	Inadequacy in design in Employer's Requirements.
2.14.2	Discrepancy in the Employer's Requirements.
2.15.2	Change in statutory requirements or decision of the development control authority after the base date.
2.18	Fees legally demandable under any statutory requirement and stated by way of a provisional sum.
2.20	Patent rights.
2.35	Defects not to be made good.
3.6	Non-compliance with instructions.
3.9	Instructions requiring changes.
3.12	Inspection – tests.
3.13.3	If tests are not reasonable.
3.14	If instructions not reasonably necessary.
3.17	Loss and expense arising from finding antiquities.
4.2	Items included in adjustments.
4.13	Ascertainment – alternative A.
4.14	Ascertainment – alternative B.
4.19	Loss and/or expense.
4.21	Addition to the contract sum.
5	Changes.
6.15.2	Remedial measures after breach of the Joint Fire Code.
6.16	Amendments to the Joint Fire Code.
Schedule 2, para. 2.1.2	If contractor is unable to enter into a sub-contract with a named sub-contractor.

Fig. 10.1 Contractual provisions regarding adjustment of the contract sum (including adjustments in the schedules).

Schedule 2, para. 2.1.6	Completion of the work of a named sub-contractor whose employment has been terminated.
Schedule 2, para. 3.2	Errors in bills of quantities.
Schedule 2, para. 4.4	Agreement of estimates.
Schedule 2, para. 4.5	Failure to agree estimates.
Schedule 2, para. 4.6	If contractor is in breach of paragraph S4.2.
Schedule 2, para. 5.5	Direct loss and/or expense.
Schedule 3, para. A.5.1	Terrorism cover – premium charge.
Schedule 3, para. B.2.1.2	If employer defaults in taking out insurance.
Schedule 3, para. B.3.5	Restoration and repair.
Schedule 3 para. C.3.1.3	If employer defaults in taking out insurance.
Schedule 3 para. C.4.5.2	Restoration and repair.
Schedule 7	Fluctuation options A, B and C.

Fig. 10.1 *Continued*

bear the cost of the error itself. This type of procurement is very unforgiving in that respect and the employer and professional advisors are unlikely to discover such errors before acceptance of the tender or indeed at any time, because of the nature of the procurement system. If, however, the employer discovers an error which is fundamental to the terms of the contractor's offer, such as the accidental omission of the fluctuations clause, the employer may not accept the offer so as to create a binding contract without drawing the error to the attention of the contractor: *McMaster University* v. *Wilchar Construction Ltd* (1971).

10.2 Interim payments

A significant difference between this form of contract and others in the JCT series is that there is no provision for the quantity surveyor to value work carried out by the contractor and there is no provision for certification by an architect of money due to the contractor. The scheme of payments is basically quite simple. Clause 4.7

states that interim payments must be made by the employer to the contractor in accordance with clause 4 and either alternative A or B, as stated in the contract particulars. The amount in the interim payment must be the gross valuation under clause 4.13 or clause 4.14 as applicable to A or B, less only the retention, any amount of advance payment due for reimbursement (see section 10.7 below) and any previous payments. Alternative A refers to stage payments and alternative B refers to periodic payments. They are fundamentally different, as explained below.

10.2.1 Stage payments – alternative A

Where the employer wishes payment to be made in accordance with a series of stages, this must be stated in the Employer's Requirements. The printed form contains a table in the contract particulars which the employer is to complete. It is a very simple table consisting of a column in which a brief description of each stage is to be inserted, and another column in which the appropriate cumulative value of the stages is to be inserted. The cumulative value of the final stage is to be equal to the contract sum. Although the descriptions of the stages are to be brief, they must be clear enough so as to positively identify the stage in question. Thus, if isolated blocks are to be the stages, it will probably be enough to note 'block 1, block 2, block 3', etc. and a final stage for the external works. If, however, the stages are physically connected, it is vital that the description precludes any dispute over the precise point at which it can be said that any particular stage has been completed. Although it may seem self-evident, periods of time are not to be inserted as stages. Thus '4 weeks, 8 weeks, 12 weeks', etc. will certainly cause the contractor to apply for the stipulated amount after the appropriate lapse of time irrespective of whether it has actually completed an adequate amount of work. The employer could have difficulty in avoiding payment in such circumstances. Instances where the stages have been completed in this way do occur and disputes have resulted.

In the previous edition of this form, it was open to the employer to include in the cumulative value the value of materials or goods stored off site. In such an instance, note was to be made in the stages column and the value of the materials and goods had to be included in the cumulative values. The first edition of this book criticised this arrangement, because it was difficult to envisage how a stage could be complete if some of the materials were stored off site. It was assumed that it was intended that the value would include the value of a completed stage plus the value of off-site materials stored for a future stage. Thus, a particular payment would include, in effect, a part payment of a future stage. This was workable, just, because the cumulative nature of the stages ensured that the payment for off-site materials was absorbed into future stage payments. This was a particular problem which could occur. In order for the employer to include the value of off-site materials in a cumulative value, some estimate of the amount of materials which would be so stored at the time that the value column was completed must have been made. When the project was in progress, the materials might not actually be delivered as scheduled. This would have left the employer in the position of being liable to pay the appropriate cumulative value which would have included the value of materials not then stored off site.

To overcome this difficulty, the employer would have been obliged to operate the deduction provisions of clause 4.10 (see section 10.4 below). That was an inelegant solution but the only one which could be devised to cope with the situation. By removing off-site materials from the cumulative value and inserting the amount into the list of amounts to be aggregated, this problem has been overcome.

Off-site materials and goods are now dealt with in clause 4.15. All such materials must be listed by the employer and the list must be supplied to the contractor and attached to the Employer's Requirements. The listed items are divided into two categories:

- Uniquely identified items, such as lift equipment, purpose made furniture or special boilers; and
- Items which are not uniquely identified, such as bricks, blocks, timber, tiles, etc.

The employer is not obliged to pay for any materials stored off site unless they are listed. Fortunately, the employer can point to the lengthy proviso which lays down specific conditions which must be fulfilled before payment for off-site materials can be included. They are:

- *The contractor must provide reasonable proof that it owns the uniquely identified items and the contractor must have provided a bond in favour of the employer if so required in the contract particulars.* The contractor may be hard pressed to provide reasonable proof in some instances. It appears that sight of the appropriate supply or sub-contract terms together with proof of payment by the contractor would be acceptable for the ownership part of this condition. The employer may decide whether a bond is required, but if so, it must be from a surety approved by the employer and the terms should be those agreed between JCT and the British Bankers' Association and set out in schedule 6, part 2.
- *The contractor must provide reasonable proof that it owns items which are not uniquely identified and the contractor must have provided a bond in favour of the employer as noted in the contract particulars.* The contractor may again be hard pressed to provide reasonable proof. Sight of the appropriate supply or sub-contract terms together with proof of payment by the contractor would probably be acceptable for the ownership part of this condition. The bond is required and it must be from a surety approved by the employer and the terms should be those agreed between JCT and the British Bankers' Association and set out in schedule 6, part 2.
- *The materials and goods must be, and remain, set apart at the place of storage or they must be clearly and visibly marked, individually or in sets by letters, figures or by reference to a pre-determined code so as to identify the employer and the destination must be marked as being the Works.* The main purpose of this condition is to make clear that these particular goods have been set aside as the property of the employer. In the case of a liquidation of the main contractor or another firm on whose premises they are stored, the employer would be entitled to recover whatever items could be positively identified as belonging to the employer. The employer would have no chance of recovery, for example, despite having paid for two dozen sink units, if none of the 50 sink units in the contractor's store were clearly

marked or set aside as the employer's property. Even marking is not foolproof. The marking necessarily has to be temporary in character and it is not unknown for an unscrupulous contractor to affix labels to a set of sanitary fittings until they had been inspected for payment on one job, then to replace them with a different set of labels to obtain payment under a quite different contract.

- *The contractor must provide reasonable proof that the listed items are fully insured for their full value under a policy protecting both employer and contractor in respect of specified perils. The insurance must cover the period from transfer of ownership to the contractor until they are delivered to the site.* This kind of proof can be discharged by giving the employer sight of the policy documents and premium receipts.

Clause 2.22 states that when the value has been included in an interim payment, the materials and goods become the property of the employer. After that, the contractor must not remove them from the premises except for use on site. The contractor, however, retains liability for damage or loss, the cost of storage, handling and insurance until they are delivered to site.

The gross valuation for the purpose of interim payments is separated into amounts which are and amounts which are not subject to retention.

Clause 4.13.1 specifies the following amounts which are subject to retention:

- The cumulative value at the appropriate stage. This is the amount noted in the contract particulars 4.7, alternative A opposite the completed stage.
- The amounts of valuations of changes or of instructions regarding the expenditure of provisional sums which are referable to the interim payment. For example, if the stages are given as 'block 1, block 2', etc., the amounts must be in respect of changes to Requirements or expenditure of provisional sums in respect of the particular block. The employer must take care that if changes are required, the instruction separates the changes referable to each block. Where provisional sums are included in the Employer's Requirements, they too must be referable to particular stages. If one item is concerned which is involved in each stage, for example the heating installation, the provisional sum should be split into a number of provisional sums. The valuation of restoration, replacement or repair of loss or damage and removal and disposal of debris which are to be treated under schedule 3, paragraphs B.3.5 and C.4.5.2 as if they were a change in the Employer's Requirements, is excluded. This somewhat unwieldy provision is intended to ensure that retention is not deducted from the value of this restoration work, because retention will already have been deducted when the work was first carried out. This item, therefore, is included in the section below.
- The total value of listed items of materials and goods stored off site which satisfy the provisions of clause 4.15.
- The amount of any adjustment made under fluctuations option C (schedule 7).

Clause 4.13.2 specifies the following amounts not subject to retention:

- Amounts resulting from payments made to or costs incurred by the contractor due to early use by the employer, patent rights, instructions regarding opening up and testing, terrorism cover under paragraph A.5.1 or taking out insurance on the employer's default in paragraphs B.2.1.2 or C.3.1 (schedule 3).

- Amounts in respect of loss and/or expense under clause 4.19 or 3.17 (antiquities).
- Amounts noted above in respect of restoration, repair, etc. treated in paragraphs B.3.5 and C.4.5.2 in schedule 3 as if they were changes in the Employer's Requirements.
- Amounts payable to the contractor under fluctuation options A (contribution, levy and tax fluctuations) or B (labour and materials cost and tax fluctuations) of schedule 7.

Clause 4.13.3 sets the amounts to be deducted:

- Appropriate deductions following instructions that defects are not to be made good during or after the rectification period or following non-compliance with instructions.
- Amounts allowable under the fluctuation options A (contribution, levy and tax fluctuations) or B (labour and materials costs and tax fluctuations) of schedule 7.

10.2.2 Periodic payments – alternative B

An employer who wishes payment to be made by periodic payments must complete the contract particulars 4.7, alternative B. If no period is stated, the period between applications for payment will be 1 month and the first interim payment must be made within 1 month of the date of possession. Since the contrary is not stated, this is a calendar month. It is open to the employer to allow the inclusion of the value of materials and goods off site in periodic payments. It is important that this information be included in the Employer's Requirements to allow the contractor to take it into account when formulating its Proposals. Where the value of off-site materials and goods are to be included, such inclusion is subject to the attachment of a list and the satisfaction of the appropriate conditions noted above in connection with off-site values in stage payments. Once again, the gross valuation for the purpose of interim payments is separated into amounts subject or not subject to retention.

Clause 4.14.1 specifies that the following are subject to retention:

- The total value of work, including design work, properly executed, including valuations of changes and expenditure of provisional sums, any adjustment under fluctuations option C of schedule 7 (use of price adjustment formulae) if applicable, but excluding the valuation of restoration etc. under paragraphs B.3.5 and C.4.5.2 of schedule 3 as noted above under the similar provision for stage payments.
- The total value of materials and goods delivered to site and intended for incorporation in the Works. There is a proviso that the goods must not be delivered prematurely and they must be adequately protected against the weather and other things.
- The total value of listed items of materials and goods stored off site which satisfy the provisions of clause 4.15.

Clause 4.14.2 stipulates the following amounts which are not subject to retention:

- Amounts resulting from payments made or costs incurred by the contractor due to early use by the employer, patent rights, instructions regarding opening up and testing, taking out insurance on the employer's default in paragraphs B.2.1.2 or C.3.1 (schedule 3).
- Amounts in respect of loss and/or expense under clause 4.19 or 3.17 (antiquities).
- Amounts noted above in respect of restoration, repair, etc. treated in paragraphs B.3.5 and C.4.5.2 of schedule 3 as if they were changes in the Employer's Requirements.
- Amounts payable to the contractor under the fluctuation options A (contribution, levy and tax fluctuations) and B (labour and materials cost and tax fluctuations) of schedule 7.

Clause 4.14.3 sets out the amounts to be deducted. These amounts are not subject to retention:

- Appropriate deductions following instructions that defects are not to be made good during or after the rectification period or following non-compliance with instructions.
- Amounts allowable under the fluctuation options A or B of schedule 7, if applicable.

10.3 Applications

It is for the contractor to apply to the employer for payment. To that extent the contractor is in the driving seat. The scheme of the contract is that the contractor carries out the valuation described in sections 10.2.1 or 10.2.2 above, as appropriate. After that, it is for the employer to pay unless the amount is disputed. Clause 4.10.1 provides that the employer has 14 days from the date of receipt of each interim application in which to pay the contractor. Applications are dealt with in clause 4.9.

If it is stated in the contract particulars that payment is to be made by stages, the contractor must make application for interim payment on completion of each stage. The contractor may make one further interim application which is to be after the end of the rectification period or on the issue of the notice of completion of making good, whichever is later. This is to deal with the release of the second half of the retention. Something which may cause difficulty is that 'completion' of a stage is not defined. It is probably sensible to consider 'completion' as very much the same kind of condition as 'practical completion' of the project as a whole, but applied to the particular stage described in the schedule. When considering the meaning of 'completion' in clause 27.4.4 of JCT 80 in *Emson Eastern Ltd* v. *E M E Developments* (1991), the judge said:

> 'In my opinion, there is no room for "completion" as distinct from "practical completion". Because a building can seldom, if ever, be built precisely as required by drawings and specification, the contract realistically refers to "practical completion" and not "completion", but they mean the same.'

The only problem which is likely to arise with regard to the last interim application is that the employer may be dilatory in issuing the notice of completion of making good. In no sense is the notice intended to be a certificate such as the architect issues under SBC and it is notable that the word 'certificate' has been avoided. Nevertheless, it is thought that the right to make application depends on it. A contractor faced with the employer's failure to issue should make a formal request for such notice followed by notice of adjudication on the matter. It should be capable of very quick resolution. It is not something which entitles the contractor to terminate its employment under clause 8.9 (see Chapter 12), because the contractor's right to payment does not arise until a proper application under clause 4.9 has been made and payment withheld without notice.

If it is stated in the contract particulars that periodic payments apply, the contractor must make application at whatever periods are stated against clause 4.7, alternative B. Note that this is a duty. The contractor is obliged to make application as prescribed. For example, the contractor may not forego making application because it wants to keep its income low in a particular period for tax purposes. The application must be made and then the employer has a good idea, from month to month, just how the finance is working. Were it not for this duty placed on a contractor, it could, for its own purposes, virtually make application to suit its own cash requirements.

The starting date should be set out for the applications in the contract particulars. It is suggested that, if 1 month is the period in alternative B, the first application should be made 1 month after the contractor takes possession of the site. If this part of the contract particulars is not completed, the default position is that applications are to be made at intervals which must not exceed 1 month, up to the date of practical completion or to a date within 1 month thereafter. However, the default position continues to state that the first interim payment (note, not the application for payment) is to be made within 1 month of the date of possession. In order to achieve that, the contractor's first application must be made at least 14 days before the end of the month following the date of possession. Applications must continue to be made until the end of the period in which practical completion occurs. For example, and assuming that the contract particulars have been completed, if the date for possession is 3 September 1999 and the date of practical completion is 2 May 2000, the first application would be made on 3 October 1999 and the last regular application would be made on 3 May 2000. After practical completion, applications must be made, not at regular intervals but when further amounts are due, particularly after the end of the rectification period or on the issue of the notice of completion of making good, whichever is later. There is an overriding proviso that the employer may not be required to make payment within one calendar month (whatever the period for interim applications may be) of having made a previous interim payment.

Clause 4.9.3 makes the important proviso that each application must be accompanied by whatever details the employer has set out in the Requirements. This applies whether the payments are to be by stages or periodic. The employer will certainly wish to check the contractor's application before payment and this will usually be done through the medium of the agent, but if the agent is an architect or an engineer, the employer may employ a quantity surveyor for the purpose depending on the size and complexity of the project. It is essential, therefore, that the details requested in the Employer's Requirements are such as will assist the employer or the agent in checking the application. In the absence of any precise specification of such details in the Employer's Requirements, it is considered that the contractor would be entitled to submit its application as a lump sum, which would be almost impossible to check. Applications for stage payments are easiest to check even though the contract provides that the actual figure applied for may be somewhat more (or less) than set out in the stages (see section 10.2.1 above). Correspondingly greater thought must be given to the calculation of the amounts of stage payments before the contract is executed.

All construction contracts must comply with the Housing Grants, Construction and Regeneration Act 1996. The payment provisions do not appear to comply with the Act in one, quite important, respect. Section 110(1)(a) of the Act stipulates:

> 'Every construction contract shall provide an adequate mechanism for determining what payments become due under the contract, and when . . .'

There is no doubt that the contract provides a means of working out 'what' becomes due, but there is nothing to say 'when' it becomes due. The Act goes on to refer to the 'date on which payment becomes due' and the period between that date and the final date for payment (which the parties are at liberty to agree). It cannot simply be assumed that the date of receipt of the contractor's application is the date payment is due, because the contract does not so state. Moreover, the Act stipulates in section 110(2):

> 'Every construction contract shall provide for the giving of notice by a party not later than 5 days after the date on which a payment becomes due from him under the contract . . . specifying the amount (if any) of the payment made or proposed to be made, and the basis on which that amount was calculated.'

Presumably, it is this part of the Act which clause 4.10.3 is intended to cover. However, under the provisions of clause 4.10.3, the employer must give the written notice not later than 5 days after receipt of the contractor's application not, as required by the Act, from the date payment becomes due (which of course the contract does not indicate). To the extent that the contract does not contain the provisions set out in section 110, section 110(3) states that the relevant provisions of the Scheme for Construction Contracts (England and Wales) Regulations 1998 (the 'Scheme') apply. The relevant part of the Scheme is part II, paragraph 4 which states:

> 'Any payment of a kind mentioned in paragraph 2 above shall become due on whichever of the following dates occurs later –
>
> (a) the expiry of 7 days following the relevant period mentioned in paragraph 2(1) above, or
> (b) the making of a claim by the payee.'

The logical and, so it appears, the legally correct conclusion is that:

- The contract does not state the date that payment becomes due.
- Therefore, paragraph 4 of the Scheme applies.

- Therefore, if application is made by the contractor on the correct date in accordance with the contract particulars, the date that payment becomes due will be 7 days thereafter.
- But if application is made by the contractor more than 7 days after the correct date in accordance with the contract particulars, the date that payment becomes due will be immediately the application is received.
- That is because paragraph 4 refers to whichever date is later.

Therefore, the date payment becomes due under the Scheme may not be the same as the date the contractor's application is received by the employer; it may be up to 7 days later. It follows that compliance with the contract clause 4.10.3 may result in the notice being served before the date payment becomes due. Section 110((2) (quoted in part above) states that the notice is to be sent *after* the date payment becomes due. In some cases, that might result in the notice being given up to 7 + 5 days after the date of receipt of the application. This is a confusing situation which it is to be hoped the JCT will address very quickly. In the meantime, no doubt employers will continue to give clause 4.10.3 notices within 5 days of the date of receipt of the contractor's application and, in an appropriate case, the contractor may challenge the notice as being too early and, therefore, a nullity.

The date payment becomes due also immediately raises the question whether the employer is in breach of contract for failure to pay when payment is 'due'. The Act's concept of a date on which payment is due and a completely different and later date which is the final date for payment, is not easy to grasp. Even if the interpretation given above is accepted, the contract gives redress to the contractor in terms of interest payment, suspension of performance of obligations and termination only if the employer fails to pay by the final date for payment. Although it may not be surprising that this contract imposes no sanctions for failure to pay by the date payment is due (because such date is not referred to), other construction contracts, which do refer to the date payment becomes due, do not impose sanctions either until after the final date for payment. Some of the problems originate from the way the Act is drafted.

The clause 4.10.3 notice must specify the amount of payment the employer proposes to make, to what it relates and the basis of calculation. It is probably of little consequence if the employer forgets to give this notice provided payment is made in full by the final date for payment. If the employer fails to pay by the final date for payment, failure to give this notice might well have serious consequences on the right to withhold payment unless the notice under clause 4.10.4 (see below) falls into the same time frame. Although this would be implied from the operation of the Act, it is expressly stated in clause 4.10.5.

Under WCD 98, a failure on the part of the employer to issue the relevant notices at the right time gave the contractor the right to payment of the amount in its application and there was no defence against the contractor's claim: *Watkin Jones & Son Ltd* v. *Lidl UK GmbH* (2002). The situation under DB, while still serious for the employer, is not hopeless. Clause 4.10.5 now provides that, if no withholding notice is given under clause 4.10.4, the amount to be paid is the amount in the clause 4.10.3 notice (the amount proposed to be paid). If no such notice has been given, the amount to be paid is to be calculated in accordance with clause 4.8. The contract does not state who is to carry out the calculation, but in the first instance obviously it will be the employer. The contractor could challenge the payment on the grounds that it does not comply with the application, but unlike the position under WCD 98, the contractor may not be successful.

If the employer fails to pay any amount due to the contractor by the final date for interim payment (i.e. 14 days from the date of receipt by the employer of the contractor's application for payment), clause 4.10.6 requires the employer to pay simple interest on the outstanding amount at the rate of 5% above Bank of England current base rate. It is to be treated as a debt. In other words, the contractor can bring an action for recovery. The clause makes clear that payment of the interest is not to be taken as a waiver of the contractor's other rights, namely its right to proper payment of the principal owing at the right time, or its right to suspend performance of its obligations, or its right to terminate its employment. It is another contractual remedy for late payment and it can be exercised whether or not the contractor also opts to suspend or determine. An employer in financial difficulties is not to treat this provision as a kind of extended loan arrangement albeit with high interest rates.

10.4 Employer's right to withhold payment or to deduct

The employer's contractual right of withholding, or deduction from, money due to the contractor is contained in clause 4.10.4. The employer may exercise any right under the contract against any amount due to the contractor even if retention over which the employer is trustee is included in the amount (clause 4.10.2). If the employer considers that he or she is entitled to withhold or deduct an amount from a payment due to the contractor, the employer must give a written notice to the contractor not later than 5 days before the final date for payment. The notice cannot be served earlier than the application to which it relates: *Strathmore Building Services Ltd* v. *Colin Scott Greig t/a Hestia Fireside Designs* (2001). The notice must state the grounds for withholding payment and the amount of money withheld on each of the grounds. The information must be detailed enough to allow the contractor to understand the reason why it is not receiving the amount withheld. Of course, the contractor may seek immediate adjudication under clause 9.2 (see Chapter 13). The contract gives the employer the right of deduction under the following clauses:

6.4.3	The cost of insurance premiums following the contractor's failure to insure.
6.5.3	The cost of insurance premiums following the contractor's failure to insure.
Schedule 3, paragraph A.2	The cost of insurance premiums following the contractor's failure to insure.
2.29	Liquidated damages.

Some other instances, such as the cost of employing others following non-compliance with instructions or the appropriate deduction after instructing the contractor not to make good defects, which used to be dealt with in this way under WCD 98 are now to be deducted from the contract sum.

In addition to the employer's contractual right of deduction, there is also the common law right of set-off, which can only be excluded by an express clause to that effect: *Gilbert Ash (Northern) Ltd* v. *Modern Engineering (Bristol) Ltd* (1973). Although at one time it was rare for an employer to exercise this right, it is becoming more common, certainly under traditional contract forms, but the employer must have a genuine reason for set-off to succeed (*C M Pillings & Co Ltd* v. *Kent Investments Ltd* (1985); *R M Douglas Construction Ltd* v. *Bass Leisure Ltd* (1991)), and the employer is obliged to comply with clause 4.10.4 and serve the relevant withholding notice.

Clause 4.10.4 is important from the employer's point of view. It provides a deceptively simple machinery for the employer to withhold payment and it applies only to interim payments. It is triggered by receipt of the contractor's application for an interim payment. It is not good enough for the employer simply to say in the notice that some of the work included in the application is defective. The employer must give a reasonable description and quantification of the sum withheld so that the contractor can consider its position.

Whether the amount stated in the contractor's application or the payment made by the employer is in accordance with the contract can only be decided by looking at the contract provisions in any particular circumstance. The onus appears to be on the employer to show in the notice that the amount claimed is not in accordance. The likely reason why an amount is not in accordance with the contract will be because it does not comply with the payment provision in clause 4. It could be, for example, that the employer disputes that a stage has been completed, or contends that materials delivered to site are too early. Experience suggests that the major reason given by the employer for withholding payment will be because it is alleged that there are defects in the work. In such a case, the employer would have to demonstrate that the workmanship and/or materials did not comply with clause 2.2. It would be helpful to that contention if, before the application, the employer has used powers under clause 3.13 in respect of the alleged defects. In the past, many employers simply deducted round figures vaguely referenced to 'defects in the work'. An employer behaving in this manner is asking for, and deserves, a notice of termination. The intention of these clauses is clear. It gives the contractor a leading role in obtaining payment, but it provides the employer with a remedy within strict procedures if the contractor tries to get more than its entitlement. Both parties should treat the whole of clause 4.10 with great respect.

10.5 Retention

The rules for ascertainment of retention are set out in clause 4.17 and they deserve careful study. The provisions are straightforward. The employer is entitled to retain a percentage of the total value of work, materials and goods included under either clause 4.13.1 (where stage payments apply) or clause 4.14.1 (where periodic payments apply). The percentage is 3% unless the parties have inserted a lower rate in the contract particulars. As the contract figure increases, the amount retained becomes progressively greater and it is a factor which the contractor has to take into account when calculating its tender. This is probably one of the reasons why the default percentage in the contract particulars is 3%. The retention is to be released in two equal parts: part one after the Works or section has reached practical completion and part two after the notice of completion of making good has been issued. If partial possession has been taken by the employer under clause

2.30, the retention is released in proportion to the value of the relevant part taken into possession and then, subsequently, a further and equal amount when notice of completion of making good of that part has been issued.

Under clause 4.16, the retention is said to be subject to certain rules. Clause 4.16.1 states that the employer's interest in the retention is fiduciary, but without obligation to invest. This clause acknowledges that the money thus retained is in reality the contractor's money which the employer is simply holding in trust for the contractor. On the face of it, this is good news for the contractor because it means that if the employer should become insolvent, the retention money should be safe and protected away from the clutches of creditors. It is well established, however, that in the event of an insolvency, such a trust fund is effective only if it is readily identifiable as belonging to the contractor. Clause 4.16.2, therefore, provides that if the contractor so requests, the employer must put aside into a separate bank account the amount of retention deducted at the date of each interim payment. The account must be identified as trust money held under the contract provisions and the employer must give the contractor written notice to that effect. The employer is entitled to the full beneficial interest and has no duty to account for it to the contractor.

The contractor's right to have the retention put into a separate account as stipulated by this clause has been supported by the courts, who have been prepared to issue mandatory injunctions for the purpose: *Rayack Construction Ltd* v. *Lampeter Meat Co Ltd* (1979). More importantly, the Court of Appeal has decided that even in the absence of an express obligation to place the retention in a separate bank account, the employer is still obliged so to do: *Wates Construction (London) Ltd* v. *Franthom Property Ltd* (1991). In that case, which concerned a very similar retention clause in JCT 98, the clause obliging the employer to put the retention in a separate bank account was deleted. The court said:

> '. . . clause 30.5.1 creates a clear trust in favour of the contractors and sub-contractors of the retention fund of which the employer is a trustee. The employer would be in breach of his trust if he hazarded the fund by using it in his business and it is his first duty to safeguard the fund in the interests of the beneficiaries . . .'.

The court had this to say about the deletion:

> 'Firstly, it seems to me that there is no ambiguity about the part of the agreement which remains. The words of clause 30.5.1 under which the trust is created are quite clear. Secondly, the fact of deletion in the present case is of no assistance because the parties, in agreeing to the deletion of clause 30.5.3, may well have had different reasons for doing so and it is not possible to draw from the deletion of that clause a settled intention of the parties common to each of them that the ordinary incidence of the duties of trustees clearly created by clause 30.5.1 were to be modified or indeed removed. It may have been thought by one of the parties to have been unnecessary to have included clause 30.5.3. It may have been that one of them thought that the employer should have been liable to account for any interest to the contractor if the retention fund was placed in a separate account. But there may be various reasons, which it is not possible to set out in full, why the clause was deleted and it is quite impossible to draw any clear inference from the fact of deletion. I therefore would reject an argument based upon the fact of deletion and can see no ambiguity upon which reference to that deleted clause could assist.'

It is established that the contractor does not have to make a request to have the retention put into a separate bank account every time an interim payment is made. Indeed, it seems that the employer has the obligation whether or not there is a clause to that effect and whether or not the contractor so requests: *Concorde Construction Co Ltd* v. *Colgan Co Ltd* (1984). The employer cannot rely on a right to deduct liquidated damages to extinguish the retention fund and so leave no obligation in respect of a separate account where WCD 98 is used: *J F Finnegan Ltd* v. *Ford Seller Morris Developments Ltd (No. 1)* (1991). The Court there held that the employer's own notice of non-completion did not have the same binding effect as did an architect's certificate issued in similar circumstances. Termination of the contractor's employment under the contract has no effect on its right to require the retention fund to be set aside. It appears, however, that there is no trust established until the separate fund has been set aside, and if the employer's insolvency predates the setting aside, the contractor will be in no better position in respect of the retention than would any other ordinary creditor: *MacJordan Construction Ltd* v. *Brookmount Erostin Ltd (in Administrative Receivership)* (1991).

There is a proviso to clause 4.16.2 which excepts local authorities from its effect. In the light of the case law noted above, it seems that the exemption is ineffective and it is surprising that the JCT have included it. If the employer wishes to avoid setting up a separate account, it may be that an appropriately worded clause in place of clauses 4.16.1 and 4.16.2 would suffice. In that case, it seems that a trust would never arise. The position is now unsatisfactory if the employer avoids setting up a separate account, and contractors should request a separate account and ensure that it is set up. For safety, the separate bank account should state:

- The name of the project; and
- That it is a trust fund; and
- That the employer is trustee; and
- That the contractor is the beneficiary.

The bank manager should be informed by letter from the employer, with a copy to the contractor, that it is a trust fund under clause 4.16.2 of the contract. If this is not done and appropriate evidence supplied to the contractor within a reasonable time, the contractor has no real alternative but to seek a mandatory injunction to enforce its rights under contract.

This book expresses no opinion about whether the clear words of the contract are sufficient to avoid what appears to be the employer's statutory duty as trustee to account for the interest to the contractor. The duties of a trustee are set out in the Trustee Act 1925 and the Trustee Investments Act 1961. No doubt the question will come before the courts in due course. By virtue of clause 4.10.2, the employer is entitled to exercise any right of deduction from money due or to become due to the contractor even if retention is part of that money.

10.6 Final payment

The final payment process is triggered by the practical completion statement under clause 2.27. Within the 3 months following, clause 4.12.1 stipulates that the contractor must submit the final account and final statement for the employer's agreement.

The contractor must supply the employer with 'such supporting documents as the Employer may reasonably require'. The contractor must supply such documents within the 3-month period, therefore it is essential that the employer informs the contractor of the requirements in good time. If the employer neglects to inform the contractor, the contractor should take the initiative by requesting details. In practice, the employer may argue that it is not possible to state what supporting documents will be required until the contractor's final account is received. Although that is certainly true so far as detail is concerned, there is nothing to prevent the employer from letting the contractor know the kind of documents required. There is nothing in this clause to suggest that the employer must state in the Employer's Requirements, the documents required. Were that intended, an express statement would have been included such as is contained in clause 4.9.3. The requirement for documents must be reasonable.

Clause 4.12.2 provides that the contract sum is to be adjusted in accordance with the clauses of the contract. The final account must show the way in which the contract sum is adjusted. Thus adjustments are set out in clause 4.2.

The contract sum must be adjusted in two important ways:

- Amounts agreed by the employer and the contractor for clause 5.1.2 changes.
- Variation in premium for terrorism cover under schedule 3, paragraph A.5.1 (if insurance option A applies).

Four categories are to be deducted:

- All provisional sums in the Employer's Requirements.
- The amount of any valuation of omissions resulting from a change, together with amounts in the Contract Sum Analysis for other work in clause 5.4.
- Any deductions resulting from clause 2.10.1 (boundary/Employer's Requirements discrepancies), clause 2.15.2 (change in statutory requirements after base date), clause 2.35 (appropriate deductions for defects not made good), clause 6.15.2 (failure to carry out remedial measures following breach of the Joint Fire Code) or clause 3.6 (non-compliance with instructions).
- Any other amounts deductible under the provisions of the contract.

Seven categories are to be added:

- Payments made or costs incurred by the contractor under clause 2.10.1 (boundary/Employer's Requirements discrepancies), clause 2.15.1 (change in statutory requirements after base date), clause 2.20 (royalty payments), clause 3.12 (inspections and testing) and clause 6.5.1 (insurance premium).
- Amounts of valuations of changes together with amounts resulting from consequential changes in conditions of work under clause 5.
- Amounts of valuation of work in accordance with instructions on the expenditure of provisional sums.
- Amounts ascertained under clause 3.17 (direct loss and/or expense following the discovery of antiquities) and clause 4.19 (direct loss and/or expense).
- Amounts paid under schedule 3, option C, after the employer's failure to maintain insurance.

- Amounts payable under the fluctuation clauses.
- Any other amounts to be added under the provisions of the contract.

The final statement must set out the amount which results after the contract sum has been adjusted and also the total amount which has already been paid to the contractor. The difference between the two amounts is to be shown as a balance payable either by the employer or by the contractor as appropriate. The statement must say to what the balance relates and the basis of calculation.

Clause 4.12.4 sets out a further timescale. The employer has 1 month to dispute the final account or final statement as submitted by the contractor. The period is measured from:

- The end of the rectification period; or
- The day named in the notice of completion of making good; or
- The date of submission of the final account and final statement by the contractor,

whichever is latest.

If the employer does not dispute it during the prescribed period, the final account and final statement are conclusive regarding the balance due to either the contractor or to the employer as appropriate. If the employer does dispute it or disputes part of it, the final account and final statement are conclusive only in respect of the part not disputed. It appears to follow that if the whole of the account and statement are disputed, none of it becomes conclusive.

It is sometimes said that, in order to dispute the final account, it is necessary to initiate adjudication, arbitration or other proceedings. That appears to be a wrong view of the position. It is clear that a dispute must arise before dispute resolution provisions are invoked; indeed, lack of a dispute will deprive the adjudicator or, as the case may be, the arbitrator of jurisdiction. In order to dispute the final account, it is simply necessary for the employer to make known to the contractor, preferably in writing, that the employer disputes or disagrees with all or part of the final account. On the plain wording of the contract, that is sufficient to prevent the final account becoming conclusive.

If the contractor fails to submit its final account and final statement within the 3-month period, the employer can take decisive action under clauses 4.12.5–4.12.7. The employer may give the contractor 2 months' written notice of intention to prepare the final account and final statement. If the contractor does not respond, the employer may proceed to prepare the documentation after the expiry of the 2 months. Essentially, the employer must go through the same process as would the contractor, deducting and adding amounts to adjust the contract sum in accordance with clause 4.2. The contract recognises that the employer will be restricted to using only the available information. The employer's final statement must set out the amounts in the employer's final account and the total amount already paid to the contractor, and must indicate the difference as payable by the employer or by the contractor as appropriate. Without the contractor's input, it is very likely that the final balance will be defective. If the employer has done the best practicable despite the lack of information, any deficiencies will be the result of the contractor's failure to act as required by the contract. The contractor is given 1 month to dispute the employer's final account and final statement by clause 4.12.7. The period is measured from:

- The end of the rectification period; or
- The day named in the notice of completion of making good; or
- The date of submission of the final account and final statement by the employer,

whichever is the latest.

If the contractor does not dispute it within the period, the employer's final account and final statement are conclusive regarding the amount due to either the contractor or to the employer as appropriate. The comments above regarding dispute or partial dispute of the contractor's final account and final statement are also applicable to this situation.

The contract is silent about the procedure to be followed where the account is disputed. Presumably, it is for the employer and the contractor to try to settle the dispute by agreeing an amount to replace the amount disputed. The dispute is something which could be referred to adjudication under clause 9.2 or to arbitration under clauses 9.3–9.8, or could be determined by legal proceedings as appropriate. The problem is that total conclusiveness is only achievable under clause 4.12.4 if the employer disputes nothing. Although the contract seems to recognise the concept of partial conclusiveness in clauses 4.12.4 and 4.12.7, other clauses do not seem to accept anything other than total conclusiveness. A number of matters hinge on conclusiveness under clauses 4.12.4 and 4.12.9. The timing of notices and, in particular, payment under clauses 4.12.8 and 4.12.9 and the conclusiveness of the final statement in respect of various matters under clause 1.9.1 are effectively now dependent on the final statement not being disputed. Once a dispute is registered, there is no defined method of gaining the conclusiveness. It is disturbing to think that the contract may have the seeds of its own frustration in these clauses.

If the final statement is prevented from being conclusive in respect of the matters set out in clause 1.9.1, it may be irritating but in most cases it will not be disastrous. Prevention of the final payment, however, would be very serious. There is no immediately obvious answer. In most cases, no doubt the parties will agree a final account in due course and they will observe the final payment provision from that point. If the employer chooses to be obstructive, however, the contract does not provide any immediately obvious solution, at least so far as can be seen. It is probable that, in order to give the contract business efficacy, a term would have to be implied to the effect that in the event of an initial dispute, the final account and final statement would be conclusive regarding the balance on the date that agreement was reached and recorded by the employer and the contractor in respect of the part or parts disputed, or failing agreement, on the date of a decision of an adjudicator, award of an arbitrator or judgment of a court as the case may be. The last thing needed in a contract form of such length and complexity is uncertainty about a significant term and JCT should deal with this point at the earliest opportunity. This problem was also present in WCD 98 and the last edition of this book drew attention to the problem in similar terms. It is difficult to understand why the JCT, in entirely re-drafting this contract in 2005, have chosen to retain this curious anomaly. The 'Design and Build Contract Guide (DB/G)' offers no explanation or indeed, guidance on this point. Meanwhile, users should consider amending these provisions.

Clause 4.12.8 stipulates that the employer must give written notice to the contractor specifying the amount of payment to be made. The notice must be given not later than 5 days after the final statement or the employer's final statement becomes conclusive about the balance due between the employer and the contractor. This notice is for the same purpose in respect of the final payment as the notice given under clause 4.10.3 in respect of applications for interim payment.

Clause 4.12.9 provides that 28 days after the final statement or the employer's final statement becomes conclusive, is the final date for payment of the balance in the final statement by one or the other of the employer or the contractor as applicable. An employer who intends to withhold or deduct payment from the final payment, must give a written notice to the contractor not later than 5 days before the final date for payment. The notice must state the amount to be withheld, the grounds for withholding and the amount to be withheld in respect of each of the grounds. This provision echoes clause 4.10.4 in regard to applications for interim payment. An employer who fails to give the notices required under clauses 4.12.8 or 4.12.9 is obliged to pay the balance, if any, stated as due in the final statement or the employer's final statement without any deduction (clause 4.12.10). This is virtually identical to the equivalent provision in WCD 98.

If the employer or the contractor fails to pay the balance or the appropriate part (after allowable deductions) to the other by the final date for payment, the party owing the money must pay simple interest at the rate of 5% over the current base rate of the Bank of England as well as the amount not properly paid. Payment of interest is not to be construed by the paying party as a waiver of the receiving party's rights to full payment (clause 4.12.11).

Clause 1.9.1 states that when the final statement or the employer's final statement has become conclusive (see the comments above) regarding the balance due between the parties, it is conclusive evidence in any proceedings arising out of or in connection with the contract whether by arbitration or litigation that:

- Where it is expressly stated in the Employer's Requirements or in an instruction that the particular quality of materials or the standards of workmanship are to be to the approval of the employer, the quality and standards are to the employer's reasonable satisfaction. This, therefore, applies only where the employer has expressly stated in the Requirements that quality or standards are to be to the employer's satisfaction. For example, the employer may want to retain control over the floor finishes or the standard of brickwork. In this instance, the final account and final statement would be conclusive that the floor finishes or brickwork were to the employer's reasonable satisfaction irrespective of whether the employer had actually taken the trouble to examine them. The employer should, therefore, take particular care not to include such clauses in the Requirements unless they are absolutely necessary and unless the employer or the employer's agent are certain to check them on site. The clause as originally drafted was tighter than the equivalent clause in JCT 80 and it is possible that it would have escaped the effects of the decision in *Crown Estates Commissioners* v. *Mowlem* (1994). The drafting of this clause has been tightened still further following *Crown Estates* to make clear that the final statement is not conclusive about the employer's reasonable satisfaction except to the limited extent expressly set out in the Employer's Requirements. It also now states that the final statement is not conclusive that any materials or workmanship comply with the contract, even the ones about which the employer is said to be reasonably satisfied. The precise result of that remains to be seen (clause 1.9.1.1).

- All extensions of time due under clause 2.25, and no more, have been given (clause 1.9.1.2).
- Reimbursement of loss and/or expense under clause 4.19 is in final settlement of all claims, whether for breach of contract, duty of care, statutory duty or otherwise, which the contractor may have and which arise out of any of the relevant matters in clause 4.20 (clause 1.9.1.3).

Proceedings are expressly stated to include adjudication, arbitration or legal proceedings. The reference to extensions of time and loss and/or expense was added because, in some instances, contractors were making very late claims in contract or in tort, long after the employer considered the final payment to have been discharged.

The conclusive effect of the final statement is subject to four exceptions. It is not conclusive:

- If there is any fraud. Fraud is the tort of deceit and may be a misrepresentation made knowingly or without belief in its truth or recklessly, careless whether it is true or false: *Derry* v. *Peek* (1889).
- If adjudication, arbitration or other proceedings have previously been commenced (clause 1.9.2). In this case, they are conclusive in respect of the specified topics after the proceedings have been concluded, but subject to the terms of any award or judgment or settlement or after 12 months from the issue of the final account and final statement if neither party has taken any further step in the proceedings, but subject to the terms of settlement of any previously disputed matters. The latter part of this exception is clearly intended to thwart a party who may serve notice of arbitration as a holding measure to prevent the final account and final statement becoming conclusive, but who concludes that it is not worthwhile taking the matter further. If it were not for this provision, service of the one notice would be sufficient to prevent the final account and final statement from ever becoming conclusive about the specified matters.
- If adjudication, arbitration or other proceedings has been commenced by either party within 28 days after the final account and final statement or employer's final account and employer's final statement would otherwise become conclusive, they are conclusive as specified except in respect of matters to which the proceedings relate (clause 1.9.3). The words 'would otherwise become conclusive' mean 'would become conclusive but for the proceedings'. Of course, where the employer or the contractor have disputed the whole of the final account under clauses 4.12.4 or 4.12.7, the final account will not become conclusive even if proceedings have not been commenced within the 28 days
- If the adjudicator gives a decision on a dispute after the date of submission of the final account or final statement or the employer's final account or employer's final statement, the employer or the contractor has 28 days, from the date on which the decision was given, to commence arbitration or legal proceedings to finally determine the dispute. This provision is to protect the right of either party to require arbitration or legal proceedings after an adjudicator's decision which, taking account of agreed extension of the statutory period, is given more than 28 days after the final account and final statement become conclusive.

A very important clause (1.10) provides that no payment made by the employer is of itself conclusive evidence that any design, works, materials or goods are in accordance with the contract except as noted above. This prevents the contractor contending that work carried out must be properly executed because it was included in an interim application which the employer paid without query. It is thought that, since the applications are cumulative (see clauses 4.13.1.1 and 4.14.1.1), the employer may rectify any overpayment of this sort by simple notice to the contractor, in accordance with clauses 4.10.3 or 4.10.4.

10.7 Advance payment

The contract provides in clause 4.6 for advance payment by the employer to the contractor where the contract particulars say that this clause applies. A note in the contract particulars states that the entry is not applicable where the employer is a local authority. It is not clear why it has been thought appropriate to exclude local authorities from the use of this clause; some local authorities do operate systems of advance payment. The idea is that the employer pays the contractor a lump sum, usually at the start of the project, and the contractor repays it over a period of months. It is for the employer to decide whether or not to operate this clause. If the employer decides to make an advance payment, the contract particulars must be completed accordingly by inserting the amount to be paid and the times and amounts for repayment. The contract particulars must also state whether the employer wants a bond. Normally, it is expected that if the employer is prepared to make advance payment, a bond will certainly be required, and if neither option is chosen, the default provision requires a bond. The bond is to be provided by a surety approved by the employer and the employer need not make payment of the advance until the bond has been provided. The terms of the bonds are to be those agreed by the British Bankers' Association and the JCT and set out in schedule 6. The date on which the employer must make the advance payment is also to be stated in the contract particulars. There will be many instances where the contractor will welcome an advance payment to enable it to fund the start of the project. In turn, the employer will doubtless expect to secure a price advantage. From the contractor's point of view, the repayment amounts should be carefully calculated so that they are amply covered by the expected interim payments.

10.8 Changes

Changes are dealt with in clause 5. It is very similar to the variation clause (clause 5) of SBC. The term 'change' is defined in clause 5.1. It means 'a change in the Employer's Requirements which makes necessary the alteration or modification of the design, quality or quantity of the Works, otherwise than such as may be reasonably necessary for the purposes of rectification pursuant to clause 3.13 . . .'. (It should be noted in passing that in the last edition of this book, it was thought that the reference to rectification must be an error, because rectification has been removed from clause 3.13 – see the comments in section 5.3.2. It is not clear why the 'error' has been retained in DB.) This broad definition of 'change' is stated to include four distinct categories:

- Addition, omission or substitution of any work.
- Alteration of the kind or standard of materials or goods.
- Removal of work carried out, materials or goods brought on site by the contractor for the Works and which are not defective.
- The imposition, addition, alteration or omission of any obligations or restrictions imposed by the employer in the Requirements or by a change in regard to access or use of the site, limitation of working space or hours or the carrying out of the work in a particular order. The employer may impose such obligations even though none were in the Employer's Requirements. The contractor has the right of reasonable objection to this kind of change under clause 3.5 (see section 5.3.1).

Clause 3.9 empowers the employer to issue instructions to produce a change in the Requirements. This is subject to the proviso that the employer may not produce a change which requires an alteration in design unless the contractor consents. This point is discussed in section 3.2. Where the contractor is also acting as CDM co-ordinator, it must notify the employer whether it has any objection under the CDM Regulations and, if it has, the employer must amend the instruction and the contractor is not obliged to comply until the amendment has been made (clause 3.9.4). No change will vitiate the contract. In one sense, this is superfluous wording because it is clear that the exercise of a power which is provided for in a contract by agreement of the parties cannot bring that contract to an end. However, the provision must be considered in relation to the first recital which briefly describes the Works. A change, or a number of changes, which resulted in the description in the first recital becoming inaccurate would certainly be beyond the power of the employer. It is thought that the limit to the employer's power would occur before that stage was reached. A change must not only be within the meaning given in clause 5.1, it must also bear some relation to the Works as described in the first recital.

Clause 3.11 provides that the employer may issue instructions in relation to provisional sums if they are in the Employer's Requirements. The employer has no power to issue any instructions if the provisional sums are included in the Contractor's Proposals – another reason why the Employer's Requirements must take precedence.

The valuation rules closely echo the equivalent rules in SBC. Clause 5.2 sets the scene by stipulating that the valuation of changes and provisional sum work is to be valued by agreement between the employer and the contractor or in accordance with clauses 5.4–5.7. Valuations, however they are made, must include allowance for addition and, perhaps strangely, for omission of relevant design work (clause 5.4.2). It is essential that the Contract Sum Analysis sets out a method of valuing design work. It must be rare for there to be an omission of design work because it is normally done so far in advance of the construction.

The provision for a price statement, which was in WCD 98, has been omitted.

If the valuation is not agreed, it must be carried out under clauses 5.4–5.7. These are essentially the valuation rules which existed in CD 81 before the concept of the contractor's price statement was introduced into WCD 98. The contract does not stipulate who is to carry out the valuation, but in view of the contractor's responsibility to make applications for payment under clause 4.9, it is clear that the contractor must also carry out valuations. This appears to give it considerable power, but valuations are to be carried out objectively. There is no question of opinion and an employer who disputes the sum may use powers under clauses 4.10.3 or 4.10.4 (see section 10.4 above), which may cause the contractor to seek immediate adjudication under clause 9.2. If the supplemental provisions are in operation, different procedures will apply (see section 10.9). There is no particular procedure in the rules for the valuation of design work which is or which becomes abortive. This is a point which might warrant some amendment of this clause or the contractor may lose substantial sums if faced with an employer who suffers from constant changes of mind.

The principal document to be consulted when applying the rules is the Contract Sum Analysis. If the supplemental provisions apply and the employer has opted to describe the Requirements in terms of bills of quantities in accordance with para. 3, the principal document becomes the bills of quantities. It is difficult to imagine an employer normally using this approach.

Omissions must be valued in accordance with the values in the Contract Sum Analysis. No adjustments are needed (clause 5.4.3).

Clause 5.4.2 provides that additional or substituted work may be valued in one of three ways:

- Work of similar character to the work in the Contract Sum Analysis: the valuation must be consistent with the values in the Contract Sum Analysis.
- Work of similar character to the work in the Contract Sum Analysis, but with changes in the conditions under which the work is carried out or changes in the quantity of work: the valuation must be consistent with the values in the Contract Sum Analysis with due allowance made for the changes.
- Where there is no work of a similar character to the work in the Contract Sum Analysis: a fair valuation must be made.

If the proper basis for fair valuation is considered to be daywork, the valuation must comprise the prime cost of the work together with percentage additions on each section of the prime cost at rates set out by the contractor in the Contract Sum Analysis. It is vital that the contractor includes such rates whether specifically requested or not. The contract spells out what is acceptable in terms of daywork (clause 5.5). It is to be calculated in accordance with the *Definition of Prime Cost of Daywork carried out under a Building Contract*, issued by the Royal Institution of Chartered Surveyors (RICS) and the Construction Federation. Alternatively, if the work is within the province of any specialist trade and there is a published agreement between the RICS and the appropriate employers' body, the prime cost is to be calculated in accordance with the definition in such agreement current at the base date. Where the contractor considers that the appropriate basis of valuation is daywork, it must submit what the contract persists in calling 'vouchers', but what everyone in construction knows are 'daywork sheets'. They must reach the employer for verification not later than the end of the week following the week in which the work was carried out and they must contain the names of workmen and the plant and materials employed, together with the time spent doing the work. There is no contractual requirement that the vouchers should be signed by the employer or the employer's agent, but such a signature is usual, signifying the employer's agreement that the vouchers are factually correct. It is not subsequently open to the employer to refuse payment on the grounds of errors in times, personnel, etc. (*Clusky* v.

Chamberlain (1995); *Inserco* v. *Honeywell* (1996)), but the employer may argue that daywork is an inappropriate basis. The procedure for submission and verification of daywork sheets is set out in the contract and, in such an instance, it has been held that even if the employer does not sign, the sheets are good evidence of what the contractor has done unless it can be shown that the sheets are incorrect: *J D M Accord Ltd* v. *Secretary of State for the Environment, Food and Rural Affairs* (2004).

Clause 5.4.4 provides that valuations must include an allowance for addition or reduction of site facilities, site administration and temporary works – the equivalent to what SBC would term 'preliminary items'.

It may happen that as a result of the contractor carrying out an instruction requiring a change or the expenditure of a provisional sum, the conditions under which other work is carried out are altered substantially. If so, the other work must be treated as if the alteration results from an instruction and it must be valued accordingly (clause 5.6). This provision echoes a similar term in SBC.

There is an overriding proviso that no allowance must be made in any valuation for the effect on regular progress of the Works or for any direct loss and/or expense for which the contractor would be reimbursed by any other provision of the contract. It is clear that a distinction is to be drawn between this clause and clause 4.19, under which the contractor will usually recover direct loss and/or expense. It is possible for the contractor to recover direct loss and/or expense under clause 5.7.2 if it can show that no other clause will cover the particular point.

Clause 5.3 provides that agreements and valuations carried out in accordance with clauses 5.2 and 5.4 are to be added to or deducted from the contract sum. This is a fairly pedantic, but necessary, provision to link with clause 4.3 so that the amount of such valuations can be included in the next interim payment after the valuation has been carried out.

10.9 Valuation of changes under the supplemental provisions

This is supplemental provision para. 4. It modifies the following clauses where they apply: clauses 5, 2.23–2.26 and 4.19–4.22 – changes, extensions of time and loss and/or expense, respectively. The idea behind the provision is that all the effects of a change instruction can be assessed and dealt with before or at the same time as the instruction is carried out. It has great similarities to the valuation of instructions under the Association of Consultant Architects' Form of Building Agreement (ACA 3). It imposes a strict discipline on both parties, but particularly on the contractor who pays a severe penalty if it forgets to operate the provision correctly. Contractors should be vigilant to check whether these provisions apply. Experience shows that both employer's agents and contractors frequently deal with the whole contract through to practical completion without realising that there are supplemental provisions.

The procedure is triggered by the issue of an instruction by the employer under clause 3.9, requiring a change. If the contractor or the employer decides that the instruction:

- Will have to be valued under clause 5.2; or
- Will result in an extension of time; or
- Will result in the ascertainment of loss and/or expense

the contractor *must* submit to the employer a set of estimates. It must do this before it carries out the instruction and no later than 14 days after the date of the instruction. The parties may agree a different time limit and if they cannot agree, the period is to be reasonable taking into consideration all the circumstances. The provisions anticipate a situation arising where the instruction is of such complexity that 14 days is not satisfactory. It is not thought that the contract reference to either contractor's or employer's opinion will cause any problems. In practice, if the contractor is of the opinion that the instruction will not have one or more of the effects noted, it is not in the employer's interests to attempt to overrule the contractor.

The contractor need not provide the estimates if the employer states in writing, at the time the instruction is issued or within 14 days after, that estimates are not required, or if the contractor puts forward a reasonable objection to any or all of the estimates. The contractor must act within 10 days of receiving the instruction and the objection may be from the contractor or from any of its sub-contractors. A reasonable objection could be that the instruction is such that the delaying effect on the completion date cannot be calculated until the instruction has been completed. In any event, such objection may be referred to the adjudicator for a decision under clause 9.2.

The estimates required from the contractor are as follows and they replace valuation and ascertainment of the instruction under clause 5.2 and 4.19 respectively:

- The value of the instruction. This is, to all intents and purposes, a quotation and it must be supported by full calculations referenced to the Contract Sum Analysis.
- Any additional resources required. It is not immediately clear why the contractor should have to include this item.
- A method statement showing how the instruction is to be carried out.
- Any extension of time required and, presumably to check errors, the new completion date.
- The amount of any loss and/or expense not included in any other estimate.

The contractor and the employer are to take all reasonable steps to agree the estimates. Once agreed, the estimates are binding. That means, for example, that if the contractor has estimated that it will need an extension of time of 2 weeks, but the effect of the instruction is actually 3 weeks, the contractor must bear the effect of its bad judgment. That will usually mean that it will have to try and accelerate at its own cost or it must pay liquidated damages for the overrun period.

This agreement is very important and it should be put in writing. If it is a matter of a straightforward agreement of the contractor's estimates by the employer, the simplest method is for the employer to write to the contractor confirming agreement to the contractor's estimate. If the agreement is to some modified version of the contractor's estimates, it is suggested that the employer should make a counter-offer, which the contractor can accept in writing. The parties have 10 days in which to reach agreement. They can, of course agree to extend this period if agreement seems to be near. However, they should take care that agreement is reached as quickly as possible or the contractor's progress may be seriously impeded. Failure to agree in 10 days presents the employer with two options:

- To instruct the contractor to comply with the instruction, stating that para. 4 will not apply. In other words, clauses 5, 2.23–2.26 and 4.19–4.22 will apply in the usual way.
- To withdraw the instruction. The employer is not liable for any costs incurred by the contractor except costs which are incurred in carrying out design work necessary for the preparation of the estimates. Such design work is to be treated as if it is the result of a change instruction.

If the contractor fails to provide estimates or fails to provide them on time, the consequences for the contractor are rather grim. The instruction is to be dealt with under clause 3.9, extensions of time are to be made under clauses 2.23–2.26 and the amount of loss and/or expense is to be ascertained under clauses 4.19–4.22, but the results of the application of clauses 5 and 4.19–4.22 will not be received by the contractor until inclusion in the final account and final statement at the end of the contract. Moreover, the contractor is not entitled to any interest or financing charges incurred prior to the issue of the final account and final statement.

10.10 Fluctuations

10.10.1 Choice of fluctuation clause

Clause 4.18 briefly provides that fluctuations are to be dealt with in accordance with whichever of three options is stated in the contract particulars to apply. It is refreshing to see that if no option is stated, option A is to apply, thus avoiding any uncertainty. The options are set out in schedule 7.

Option A: Contribution, levy and tax fluctuations, is used when minimum fluctuations are desired.
Option B: Labour and materials cost and tax fluctuations, is used where full fluctuations are intended.
Option C: Formula adjustment, is used alternatively where full fluctuations are desired.

These clauses are very similar to the equivalent clauses for use with SBC.

10.10.2 Option A: Contribution, levy and tax fluctuations

This option is similar to option B but it is limited to duties, taxes and the like and excludes fluctuations in the prices of labour of materials. This is the minimum fluctuation clause under this form of contract.

10.10.3 Option B: Labour and materials cost and tax fluctuations

This is the full fluctuations clause where it is not desired to use the formula. The clause is intricately drafted to achieve its effect. Put as simply as possible, the adjustments are considered in four categories:

- Rates of wages
- Contributions, levies and taxes
- Materials, goods, electricity and fuels
- Landfill tax.

Paragraph B.1.1 provides that the contract sum is based on the rates of wages, other emoluments and expenses payable by the contractor to or in respect of workpeople on the site or off site, but directly employed and engaged on production for the Works. The rules or decisions of the Construction Industry Joint Council or other wage-fixing body current at base date are to apply, together with any appropriate incentive scheme and the terms of the Building and Civil Engineering Annual and Public Holidays Agreements and the like. The contract sum also takes into account the rates of contribution, levy or tax payable by the contractor in its capacity as employer and includes such things as the cost of employer's liability insurance, third party insurance and holiday credits.

Paragraph B.1.2 stipulates that increases or decreases in the rates of wages or other emoluments and expenses due to alterations in the rules etc. after base date, must be paid to or allowed by the contractor together with consequential increases or decreases in such things as employer's liability insurance, contributions, levies and taxes, etc. Paragraph B.1.3 provides that in respect of persons employed on the site, but not defined as workpeople, the contractor must be paid or must allow the same amount as payable in respect of a craftsman under paragraph B.1.2. 'Workpeople' is defined in paragraph B.12.3 as persons whose rates of wages and other emoluments are governed by the relevant bodies noted in paragraph B.1.1. Paragraph B.1.5 provides for adjustment in respect of increases or decreases in reimbursement of fares covered by the rules of the appropriate wage-fixing body, and if the transport charges in the basic transport charges list are increased or decreased.

Paragraph B.2.1 provides that with regard to contributions, levies and taxes, the contract sum is based on the types and rates of contribution, levy and tax payable by the contractor in its capacity as an employer. Again, the datum is the types and rates payable at base date. 'Contributions, levies and taxes' is defined very broadly in paragraph B.2.8 as all impositions payable by the contractor as employer providing that they affect the cost to the employer of having persons in its employment. In general terms, changes or cessation in payments after base date are allowable as fluctuations.

Paragraph B.3.1 provides that the contract sum is based on market prices of materials, goods, electricity, fuels or any other solid, liquid or gas necessary for carrying out the Works, and on duty payable on waste disposal.

Paragraph B.4 provides that the contractor must incorporate provisions having the same effect if it sub-lets any of the Works. Any increase or decrease in the price payable under the sub-contract and which is due to the operation of such incorporated provisions is to be paid to or allowed by the contractor.

Paragraphs B.5 to B.12 are mainly concerned with procedural matters and definitions. Importantly, certain notices are to be given by the contractor. Each notice is expressed as being a condition precedent to payment being made to the contractor in respect of the event of which notice is to be given. Paragraph B.7 deals with adjustments. The contract sum may be adjusted and adjustments under option B must be taken into account when the termination provisions are implemented. The contractor is to provide such evidence and computations as the employer may reasonably require to enable the amounts to be ascertained. That is not to say that the employer is responsible for calculating fluctuations. Indeed, paragraph B.6 states that the parties may agree on the matter. In practice, the contractor will normally carry out the appropriate calculations and the Employer's Requirements should require the inclusion of all relevant calculations and evidence at the time of submission of applications for payment. Paragraph B.10 states that if the contractor is in default as regards time, recovery of fluctuations is frozen when it becomes in default. This is subject to two very important provisos:

- There must be no amendment to the printed text of clauses 2.23–2.26 (extensions of time).
- The employer must respond to every written notice of delay from the contractor by fixing a new completion date or confirming the old one. This must be done in writing.

These are points which can easily be overlooked when the contract is being set up or during the progress of the work. These fluctuation provisions are expressly not to apply to daywork rates or to changes in rate of VAT.

10.10.4 Option C: Formula adjustment

This option is used where full fluctuations are to be dealt with by the use of formulae. Adjustment is to be made in accordance with this option and the Formula Rules for use with option C issued by the JCT and current at the base date. If this system is to be used, it is essential that the Employer's Requirements request the Contract Sum Analysis in the proper form. Helpful guidance is given in Practice Note 23.

10.11 VAT

Clause 4.4 deals with value added tax and it is similar to clause 4.6 of SBC. The 'VAT Agreement' which used to be annexed to the contract is now omitted. The contract sum is exclusive of VAT and the contractor is entitled to recover from the employer any VAT which it has to pay on goods or services.

Chapter 11
Insurance and Indemnities

11.1 Injury to persons and property

The indemnity and insurance provisions were extensively revised by Amendment 1 issued in November 1986 and again to a limited extent by amendment 10 in July 1996. They are now almost identical to the insurance and indemnity clauses in SBC. For some reason best known to JCT, the insurance provisions are contained in clause 6, but the Works insurance provisions are now in schedule 3 at the back of the contract. This is already a source of confusion to users of this and other JCT contracts. Where the employer employs an agent, it will be the agent's duty either to check the contractor's insurance for the employer, or to get an expert to do so or to make sure that the employer seeks advice from an expert: *Pozzolanic Lytag Ltd* v. *Bryan Hobson Associates* (1999). The forerunner to SB (WCD 98) contained a provision in clause 22D to allow the employer to insure against loss of liquidated damages. The provision was not always properly understood and, perhaps as a consequence, not much used apparently. The provision has been omitted from DB.

Under clauses 6.1 and 6.2 the contractor assumes liability for and indemnifies the employer against any liability arising out of the execution of the Works in respect of the following:

- Personal injury or death of any person except to the extent that the injury or death is due to any act or neglect of the employer or any of the employer's persons. This will include directly employed contractors under the provisions of clause 2.6. The employer's agent will also be included. In the case of personal injury or death, therefore, the contractor is entirely liable unless some or all of the blame can be laid at the employer's door. In that case, the contractor's liability is to be reduced in proportion to the employer's liability. Thus, if the employer is 30% to blame, the contractor will be liable for the remaining 70%.
- Injury or damage to any property to the extent that the injury or damage is due to negligence, breach of statutory duty, omission or default of the contractor or any contractor's person. In this case, the contractor is not liable unless it can be shown to be at fault and even then, it is liable only to the extent it can be so demonstrated. The contractor's liability does not apply to loss or damage caused by specified perils to property for which option C, paragraph C.1 insurance has been taken out by the employer. Neither does it apply to the Works or materials on site up to the date of practical completion, section completion or partial possession of a particular part or termination of the contractor's employment.

There is a striking difference between the two indemnities. In the first one, the contractor is liable unless the employer is shown to be liable. In the second case, the contractor is only liable so far as it can be shown that the contractor is in default. The indemnity clauses are important, because they establish the occurrences for which the contractor must assume liability or partial liability. Indemnity clauses are always strictly interpreted by the courts. In particular, it requires very clear words for a person to be indemnified against the consequences of their own negligence: *Walters* v. *Whessoe* (1960). It is not thought that these clauses fall into that category.

Clause 6.4 provides that the contractor must take out and maintain insurance against its liabilities under clauses 6.1 and 6.2. This duty is stated to be without prejudice to the contractor's obligation under those clauses. Thus, the taking out of insurance does not affect the contractor's liabilities. If the contractor fails to take out or maintain the cover or if the insurer refuses to meet a claim, the contractor will have to find the money itself. The insurance must be for a sum of money which is not less than the sum stated in the contract particulars for any one occurrence or series of occurrences arising out of one event.

Where the insurance refers to the personal injury or death of one of the contractor's employees arising in the course of employment, it must comply with all relevant legislation.

The employer has the right to require the contractor to provide evidence that the insurances have been taken out and are being maintained. The only stipulation is that the request must be reasonable. In addition, the employer may ask for the policies and premium receipts at any time provided that the request is neither unreasonable nor vexatious. It is not entirely clear what the JCT had in mind when drafting that particular provision, which still seems to be repetitive in DB as it was in WCD 98. The employer is given important powers by clause 6.4.3. If the contractor does not take out or maintain the appropriate insurances under clause 6.4.1, the employer may take out the insurance, deducting the premium cost from any money due to the contractor after first serving the appropriate notice, or it can be recovered as a debt. Before taking this step, the employer must establish that the contractor is in default. Although the contract does not specify any time by which the contractor must have effected insurance, in the context of the contract as a whole, the employer and/or the contractor will be in considerable financial danger if the insurances are not in place before the contractor takes possession of the site. A wise employer will ensure that requests for details of insurances from the contractor under clause 6.4 are made before possession takes place.

11.2 Employer's liability

Clause 6.5 provides for insurance against damage caused by the carrying out of the Works, when neither contractor nor employer are negligent or in default in any way. This loophole was highlighted in *Gold* v. *Patman & Fotheringham Ltd* (1958), following which the predecessor of this clause was introduced into earlier standard forms. The clause is operative only where it is stated in the Employer's Requirements that the insurance is required. It is not clear why this statement is not included in the contract particulars as is the case with SBC, particularly as the amount of indemnity is to be stated there. If the insurance is required, the contractor must take out and maintain the insurance in joint names. A footnote states that the expiry date should not be before the end of the rectification period. By taking out insurance in joint names, both contractor and employer are the insured. This is important in respect of this type of insurance, because the employer may often be the one to receive a claim following damage to other property. If the contractor simply took out this insurance in its own name, the insurer would not be liable to cover any loss unless the claim was made against the contractor. In the broader context of the later insurance provisions, insuring in joint names ensures that if the loss is due to some negligence of either contractor or employer, the insurer cannot simply pay the insured party and then use its rights of subrogation to recover the payment from the party in default. Subrogation is the right of an insurer to stand in the place of the insured receiving payment in order to pursue legal action in the insured's name against the defaulting party.

The insurance must be taken out against any expense, liability, loss, claim or proceedings which the employer may suffer as a result of damage to any property from certain specified causes which are due to the carrying out of the Works:

- Collapse
- Subsidence
- Heave
- Vibration
- Weakening or removal of support
- Lowering of ground water.

There can be few contracts where some degree of insurance of this kind is not required, because of the lack of greenfield sites. In practice, this kind of insurance covers situations which are common in town and city centre sites. Typically, the contractor will excavate near to adjacent property, or perhaps it will be inserting piles. Although the contractor takes every reasonable precaution and carries out the work with exemplary precision, the adjacent property suffers damage because the walls are inadequately founded or a peculiarity of the underlying strata sets up a vibration which the property cannot resist. In other words, none of the parties concerned were at fault and the danger could not be anticipated – what is commonly called a pure accident. Although no one is at fault, the adjacent owner will certainly have an action against the employer and/or the contractor in the tort of nuisance. The insurance does not cover the following damage:

- If the contractor is liable under clause 6.2 (injury or damage to property real or personal), because this clause is principally for the benefit of the employer and the contractor indemnifies the employer under clause 6.2 and insures the risk under clause 6.4.
- If the damage is due to errors in the design. This presumably refers to both the design, if any, provided by the employer through professional advisors as part of the Employer's Requirements, and to the completion of the design by the contractor either directly or through its sub-contractors.
- If it is reasonably clear that damage will result from the building operations. This is a difficult exception, because very many operations in town centre sites almost inevitably cause some damage to adjacent property even if it is only slight cracking. It may depend on just what the insurers are prepared to cover after hearing expert opinion.

- If the employer must insure against just that kind of damage as part of the insurance under option C, paragraph C.1 (insurance of existing structures).
- If the damage is to the Works and materials on site.
- If the damage results from war, hostilities or the like.
- If the damage is caused by any of the excepted risks. The excepted risks are listed in clause 6.8 and cover such things as ionising radiations, nuclear risks and sonic booms.
- Caused by or arising from pollution unless occurring as a result of one sudden incident.
- If it results in the employer being liable for damages for breach of contract unless the damages would have been applicable without a contract.

Insurance under this clause must be placed with insurers to be approved by the employer. The contractor is not merely to produce the policy on request, it must deposit the policy with the employer. If the contractor defaults, the employer has power under clause 6.5.3 to take out the appropriate insurance and to deduct the cost of premiums from any sums payable to the contractor. It is essential, if this type of insurance is required, that it is taken out before the contractor takes possession of the site.

Clause 6.6 makes clear that the contractor has no liability to indemnify the employer or insure against personal injury, death or injury or damage to property including the Works and site materials caused by any of the excepted risks.

11.3 Insurance of the Works

Clause 6.7 deals with insurance of the Works and immediately refers to the insurance options in schedule 3. The scheme of insurance is relatively simple in principle although fairly complex in detail. It is virtually identical to the Works insurance provisions in SBC. There is a choice of three options:

- Option A if new buildings are to be insured by the contractor.
- Option B if new buildings are to be insured by the employer.
- Option C when the work is in existing buildings or extensions to them.

One of these clauses must apply as stated in the contract particulars. Care must be taken to choose the correct clause. For example, if option A was stated to apply when the contract called for work to an existing building, the employer may be in the dangerous position of being partially or totally uninsured and liable to foot the bill for any damage. Two categories of insurance are involved, all risks and specified perils:

- *All risks insurance*: This is a very broad category of insurance which covers any physical loss or damage to work executed and site materials, with certain exceptions set out in the clause. 'Site Materials' as noted in this definition is a term which the contract defines in clause 1.1 as 'all unfixed materials and goods delivered to, placed on or adjacent to the Works which are intended for incorporation therein'. Practice Note 22 states that the main additional risks covered by all risks insurance to those covered by the former 'clause 22 perils' are impact, subsidence, theft and vandalism. Clause 22 perils have since been replaced by specified perils.

- *Specified perils insurance*: This category of insurance is restricted to the items in the definition in clause 6.8 and includes such things as fire, flood and earthquake, but it excludes the excepted risks.

Clause 6.9 is a very important clause for sub-contractors. It places an obligation on the contractor or the employer, whoever has the duty to insure, to ensure that the joint names policies mentioned in options A, B or C have certain safeguards for any sub-contractor.

Either contractor or employer, as appropriate, must ensure that the policy either recognises each sub-contractor as an insured, or it must include a waiver of subrogation by the insurers. The protection applies only in respect of losses by the specified perils to the Works and site materials and it is to last until practical completion of the sub-contract works or termination. Either the contractor or the employer must ensure these safeguards are included in any policies which are taken out following default of the other party. The value to a sub-contractor is that, for example, a sub-contractor who causes damage to the Works by fire following negligence will not be open to any action by the insurers to recover money paid out to the employer or the contractor. This kind of protection for the sub-contractor has been held to be effective if properly reflected in the sub-contracts: *BP Exploration Operating Co Ltd* v. *Kvaerner Oilfield Products Ltd* (2004). The protection does not extend to joint names policies taken out in respect of existing structures and contents under paragraph C.1 of option C. In such cases, the insurer will retain subrogation rights and the sub-contractors owe a duty of care to the employer so as to be liable in damages if they negligently cause one of the specified perils: *British Telecommunications plc* v. *James Thomson and Sons (Engineers) Ltd* (1998). It has been held, in *Hopewell Project Management Ltd* v. *Ewbank Preece Ltd* (1998), that the protection is unlikely to extend to architects and others providing professional services to the contractor. The judge remarked that it would be most unusual for the term 'sub-contractor' in a contractor's all risks policy to include a firm which provided professional services.

11.4 Insurance of the Works: new building

The contractor's obligations are set out in schedule 3, option A, paragraph A.1. The contractor must take out and maintain a joint names policy for all risks insurance. The value must cover full reinstatement of the Works or sections and the amount of any professional fees inserted in the contract particulars. Professional fees are the fees required by the professionals involved in the reconstruction work. The employer is entitled to deduct from insurance proceeds the amount incurred for professional fees. A strict reading of the wording suggests that if the percentage is omitted from the contract particulars, the employer would be obliged to pay the fees. The reinstatement value should be carefully considered. If the building is effectively a total loss at a point when 50% of the work has been completed, the cost of demolition of what remains, together with reconstruction at inflated prices, could result in the contractor having to subsidise the project. The contractor should get very good advice from its broker before taking out insurance to cover this risk.

It does not include consequential loss: *Kruger Tissue (Industrial) Ltd* v. *Frank Galliers Ltd and DMC Industrial Roofing & Cladding Services and H & H Construction* (1998); *Horbury Building Systems Ltd* v. *Hamden Insurance NV* (2004). The insurance must be maintained until practical completion of the Works or termination of the contractor's employment, even though such termination may be the subject of dispute between the parties. The employer, who must approve the insurers, is entitled to have the policy documents and premium receipts. If the contractor defaults, the employer has the right to take out the policy and deduct the cost from any sums payable to the contractor (paragraph A.2).

In practice, of course, the contractor will usually maintain an annual policy which provides cover against all the risks which it may face. So the one policy, possibly by endorsements, will include cover against liability for injury or death to persons, injury or damage to property other than the Works, employer's liability and Works insurance. Paragraph A.3 makes provision for this situation and allows the contractor to use its annual policy to discharge its obligations under paragraph A.1 provided that the policy is in joint names (a separate endorsement is required for each contract undertaken) and that it provides cover for not less than full reinstatement and professional fees. The contractor must provide evidence that the insurance is being maintained if the employer so requests, but there is no obligation to deposit the policy. The annual renewal date is to be inserted in the contract particulars.

The mechanics of the clause are contained in paragraph A.4. If any loss or damage occurs, the contractor must give written notice to the employer as soon as it discovers the loss. A very important provision (paragraph A.4.2) makes clear that the fact that part of the Works has been damaged must be ignored when the amount payable to the contractor is being calculated. The contractor must be paid for the work carried out, although it may since have been destroyed. There should be no problems in this respect if payment is being made on a periodic basis (see section 10.2.2); the next payment application will be made and paid as usual. If payment is to be made by stages, there could be difficulties as the payments are not to be made until completion of each stage (clause 4.9.1). Thus, the contractor will be paid, but it may have to wait for some time before it can make application for payment, particularly if the damage is extensive.

Paragraph A.4.3 has been known to cause difficulties. It places a duty on the contractor to carry out restoration and remedial work and proceed with the Works when the insurers have carried out any inspection they require. It may take the insurers a considerable time to accept the claim. The contractor is not entitled to wait until it knows whether or not the claim will be accepted before it proceeds with the Works. The result is often a heavy financial burden on the contractor. If the damage is very serious, the insurers may employ their own engineers and surveyors to assess the feasibility of repair or total reconstruction. The contract makes no provision for this, but it would be an extremely foolhardy contractor that proceeded with its own ideas of reconstruction in the face of the insurers' own views. It should also be noted that a contractor is not entitled to any extension of time if the cause of the damage lies outside those items listed under specified perils. For example, if the building shell was erected and subsequently collapsed, the contractor would receive no extension of time for the resulting delay no matter who was ultimately at fault. When serious damage occurs, it is in the interests of both parties to obtain first-class advice.

The contractor and all its sub-contractors who are recognised as insured must authorise the insurers to pay insurance monies to the employer. The contractor is entitled to be paid all the money except for any percentage noted in the contract particulars for professional fees. It is thought that the effect of the wording is that the employer may retain only the amount he or she has paid out or is legally obliged to pay out in professional fees directly related to the loss or damage, but that there is a ceiling on the amount which may be retained. That ceiling is set by the percentage.

The contractor receives the insurance money from the employer in instalments in accordance with clause 4.14, alternative B (periodic payments), even if the mode of payment stipulated in the contract is actually alternative A (stage payments). The contractor is not entitled to receive more than the insurance money and, if there has been an element of underinsurance or the policy carries an excess, or if the insurers repudiate their liability, it is for the contractor to make up the shortfall.

Option B provides for new building insurance to be taken out by the employer. This is not common in practice. The obligation is principally contained in paragraph B.1 and it is similar to the contractor's duties under paragraph A.1. The employer must take out and maintain a joint names policy for all risks to cover the full reinstatement value of the Works or sections together with the percentage to cover professional fees. The employer must maintain the policy until practical completion or termination, whichever is earlier. The employer must produce evidence for the contractor that the policy has been taken out and, on default, the contractor may itself take out a similar policy and it may recover the cost as an addition to the contract sum.

Paragraph B.3 sets out the machinery for dealing with an insurance loss. It closely follows paragraph A.4 and provides for the contractor to give written notice to the employer upon discovering loss or damage. The contractor must proceed with repairs and the execution of the Works after any inspection required by the insurers, and the contractor and its sub-contractors who are noted as insured must authorise payment of insurance monies directly to the employer. Here, however, the similarity ends. Where the employer has insured, paragraph B.3.5 stipulates that restoration, replacement and repair must be treated as if they were a change in the Employer's Requirements. There are two important points to note. First, the change does not depend on an instruction from the employer. The fact that there has been loss or damage and the employer has the obligation to insure is sufficient. Second, it follows that if the repairs etc. are to be treated as a change, they are to be valued and the employer must pay for them. This duty is not affected by any shortfall or excess in the employer's insurance, nor is it affected if the insurers decide to repudiate liability. Under option B, it is the employer who must make good any shortfall.

11.5 *Insurance of the Works: existing building*

If work is to be undertaken in an existing building or in extensions to an existing building, the appropriate option is C. The insurance is to be taken out and maintained by the employer, and the employer's obligations are set out in paragraphs C.1 (existing buildings) and C.2 (Works in or extensions to existing buildings).

Paragraph C.1 refers to a policy in joint names to cover the existing building and contents for specified perils only. The contents are those which are owned by the employer or for which the employer is responsible. This is presumably intended to cover the employer's goods, goods on the premises with permission, but not goods which may be on the premises without the employer's permission. Where portions of the new Works are taken into possession by the employer under clause 2.30, they are to be considered part of the existing building from the relevant date. This is a point which the employer must watch when taking possession of portions of the Works under clause 2.30. Paragraph C.2 obliges the employer to take out a joint names policy for the new Works in respect of all risks.

Both sets of insurance must be taken out for full reinstatement value, but only in the case of the new Works must the professional fees percentage be added. Both insurances must be maintained until practical completion or termination. After that the contractor is exposed to claims for damages from the employer: *TFW Printers Ltd* v. *Interserve Project Services Ltd* (2006).

Paragraph C.3 is not to apply if the employer is a local authority. It gives the contractor broad powers if the employer defaults. The contractor has the usual power to require proof that the insurances are taken out and are being maintained. In addition, in the case of default in respect of paragraph C.1 insurance, it has right of entry into the existing premises to inspect and make a survey and an inventory of the existing structures and the contents. The only qualification on the contractor's power is that the right of entry and inspection is such as may be required to make the survey and inventory. This provision merits careful consideration by the employer, because a failure to insure by the employer may give rise to distinctly unwelcome, but lawful, entry by the contractor into the employer's property. The contractor may itself take out and maintain a joint names policy and the premium sums paid by the contractor must be added to the contract sum. Paragraph C.3.2 briefly provides that, where the employer is a local authority, copies of cover certificates must be produced by the employer on reasonable demand by the contractor to certify that terrorism cover is being provided under each policy.

The machinery for dealing with loss or damage is set out in paragraph C.4. Note that there is no express machinery for dealing with damage to the existing building and contents; the parties are left to their own devices. In the case of loss or damage to the Works, the contractor must give written notice to the employer on discovery. The contractor and its sub-contractors noted as insured must authorise the insurers to pay any insurance money directly to the employer. There is provision for either party to terminate the contractor's employment within 28 days of the occurrence if it is just and equitable to do so (see section 12.3.5).

If there is no termination or an arbitrator decides that the notice of termination should not be upheld, the procedure is much the same as paragraph B.3. The contractor is obliged to proceed after any inspection required by the insurers, but the work is to be treated as a change for which the employer must pay. Shortfalls in insurance under this clause are again the responsibility of the employer. Under none of the three Works insurance options is the contractor penalised in respect of work already carried out and damaged by the insurance risk.

It is by no means crystal clear who bears the risk in the event of damage caused to the Works or existing structures due to the contractor's negligence. Common sense appears to suggest that where damage is indisputably the result of negligence on the part of the contractor, the contractor should be responsible for the cost of remedial work. Of course, the difficulty here is that it is the employer who is responsible for the insurance. In the case of new Works insured under option A by the contractor, this problem does not exist because the contractor claims on its own insurance. The rationale behind allowing a claim on the employer's insurance for something which is the result of the contractor's negligence is that the insurance is expressly stated to be in joint names. This means that both the employer and the contractor are the insured parties and the insurers cannot exercise the right of subrogation against the contractor or the employer. There seems to be little point in this exercise if a claim cannot in any event be made against the employer's insurance where the contractor is negligent. Although there have been dissenting opinions, this view has been largely supported in the courts: *Scottish Special Housing Association* v. *Wimpey Construction UK Ltd* (1986); *British Telecommunications plc* v. *James Thompson & Sons (Engineers) Ltd* (1998) HL; *GD Construction (St Albans) Ltd* v. *Scottish & Newcastle plc* (2003) CA.

11.6 Terrorism cover

Clause 6.10 provides that if the insurers notify either the employer or the contractor that terrorism cover will stop from a specified date, the notified party must notify the other. The employer must then write to the contractor either:

- Requiring the Works to be continued; or
- Stating that on a specified date the contractor's employment will terminate. This date must be after the date of the insurer's notification, but by the cessation date of cover stated by the insurers, at the latest.

Clause 6.10.3 provides that if the option to terminate is taken by the employer, clauses 8.12.2–8.12.5 will apply, but excluding clause 8.12.3.5. Other clauses in the contract which require further payment to the contractor or release of retention must no longer apply. Clause 6.10.4 governs the situation if the employer decides not to terminate. In that instance, if the Works or site materials suffer damage due to terrorism, the contractor must restore the work and the restoration is treated as a change. Where option C applies, clause 6.10.4.3 makes clear that the employer is not obliged to reinstate any existing structure which is damaged by terrorist activity.

Insurance option A contains a provision in paragraph A.5 dealing with terrorism cover. It stipulates that the contract sum must be adjusted accordingly if the rate on which cover is based is varied when cover is renewed. There is an exception where the employer is a local authority. In such cases the employer may give instructions to the contractor that the cover is not to be renewed under the policy and, if the employer further instructs, the terms of clauses 6.10.1 and 6.10.2 will apply if the Works or site materials suffer damage by terrorism after the renewal date.

11.7 The Joint Fire Code

The Joint Fire Code is dealt with under clauses 6.13–6.16. Clause 1.1 defines the Joint Fire Code as:

'the Joint Code of Practice on the Protection from Fire of Construction Sites and Buildings Undergoing Renovation, published by the Construction Confederation and the Fire Protection Association with the support of the Association of British Insurers, the Chief and Assistant Chief Fire Officers Association and the London Fire Brigade as amended/revised from time to time.'

The code makes clear that non-compliance could result in insurance ceasing to be available. If the code is to apply, the contract particulars should be completed appropriately. If the insurer categorises the Works as a 'Large Project', special considerations apply and the contract particulars must record that also.

Clause 6.14 requires both employer and contractor and anyone employed by them, and anyone on the Works including local authorities or statutory undertakers, to comply with the code. Clause 6.15 makes clear that if there is a breach of the code and the insurer gives notice requiring remedial measures, the contractor must ensure the measures are carried out working regularly and diligently. If the contractor does not begin the remedial measures in 7 days from receipt of the notice, or if it fails to proceed regularly and diligently, the employer may employ and pay others to do the work. All additional costs incurred by the employer then become the liability of the contractor and an appropriate deduction may be made from the contract sum. If the code is amended after the base date and the contractor is put to additional cost in complying, the way the cost is to be borne is to be as stated in the contract particulars. If the employer is to bear the cost, such cost must be added to the contract sum.

11.8 Professional indemnity insurance

This is a welcome addition to the contract provisions. Clauses 6.11 and 6.12 set out the position. The provisions are quite straightforward, but there are some points to watch. Clause 6.11.1 requires the contractor to take out professional indemnity insurance forthwith (i.e. as soon as it reasonably can do so) after the contract has been entered into. It is important to remember that the contract will have been entered into as soon as the employer has unequivocally accepted a contractor's tender based on these contract terms; otherwise when the formal documents have been executed. The insurance policy is to have a limit of indemnity as set out in the contract particulars so far as the type and amount is concerned. Alternative levels of cover are stated in the contract particulars:

- Relating to claims or a series of claims arising out of one event; or
- The aggregate amount for any one period of insurance.

The note should be read carefully. If neither is selected, the amount will be the second (aggregate amount). The amount is to be inserted as a sum of money, but if no amount is inserted, a deceptively simple note states that insurance under clause 6.11 'shall not be required'. One can see the logic in inserting the note in that position, but if the employer is going to forget to insert a monetary amount, the note will not be read either. In its next amendment, the JCT might consider inserting a provision to the same effect as the note, but within clause 6.11 itself. There is also a space in the contract particulars for the insertion of a monetary amount as the level of cover for pollution or contamination claims. However, in this instance, if no amount is inserted, the level of cover is stated to be the full amount of the earlier stated indemnity cover. Clause 6.11.3 stipulates that the contractor must produce documentary evidence of insurance when reasonably requested by the employer.

Clause 6.11.2 then requires the contractor to maintain the insurance so long as it remains available at commercially reasonable rates. That is a common formula which will be familiar to most architects and other construction professionals. The insurance is to be maintained for the period stated in the contract particulars from practical completion of the Works. It is not uncommon for there to be a dispute about the date of practical completion. There is no certificate under this contract and practical completion is essentially a matter of fact to be decided, if needs be, by one of the dispute resolution procedures. The contract particulars give the option of 6 years or 12 years for the period, together with a space which can be filled in with any other period. If nothing is chosen, the default period is 6 years. Where the contract is executed as a deed, the parties will certainly be advised to choose 12 years to match the limitation period. It is quite common for the insurance period to be stated as 7 or 8 years where the contract is executed under hand, and 13 or 14 years where it is a deed, in order to make sure that the limitation period in each case is completely covered.

Clause 6.12 tries to cover the position if insurances stop being available at commercially reasonable rates. The contractor must immediately notify the employer so that they can discuss the best means of protecting their positions without insurance cover. One might easily replace the contractual words with something to the effect that if cover becomes unavailable at commercially reasonable rates, employer and contractor must get together to think what to do next. The fact is that, without insurance cover, it is difficult to think of any adequate method of protecting them. On the one hand, the employer needs to be sure that the contractor has money available to deal with any liability resulting from breach of obligation to design. On the other hand, the contractor wants the same comfort. If it is not there and the cost of rectifying a design fault is considerable, the contractor may become insolvent and the employer gets little or nothing. Failure to notify the employer as required under clause 6.12 is a breach of contract on the part of the contractor. The problem is that, if the contractor simply stops paying the premiums, the employer is not likely to know about the breach until there is a claim; then it is too late. Too minimise this danger, the employer must put in place a procedure to ensure an annual request to the contractor to provide the insurance information in accordance with clause 6.11.3.

Chapter 12
Termination

12.1 *Common law position*

Under the general law, a contract can be brought to an end in four main ways:

- By performance
- By agreement
- By frustration
- By breach and its acceptance.

12.1.1 Performance

This is the ideal way of bringing a contract to an end – when both parties have carried out their obligations under the contract and nothing further remains to be done. At that point, the purpose for which they entered into the contract has been accomplished and the contractual relationship ceases.

12.1.2 Agreement

If the parties to a contract so wish, they may agree to bring the contract to an end. What they are actually doing is entering into another contract whose sole purpose is to end the first contract. In most cases, when a contract is ended by mutual agreement it is because the parties gain something from so doing, thus satisfying the requirement for consideration as an essential element of the contract. However, it is prudent for the parties to execute the second contract as a deed, thus avoiding any question of consideration arising.

12.1.3 Frustration

A useful definition of frustration was given by Lord Radcliffe in *Davis Contractors Ltd* v. *Fareham Urban District Council* (1956):

> '(It) occurs wherever the law recognises that without default of either party a contractual obligation has become incapable of being performed because the circumstances in which performance is called for would render it a thing radically different from that which was undertaken by the contract.'

A straightforward example of a contract being frustrated is if a painting contractor entered into a contract to repaint the external woodwork of a house and before it could commence work, the house was destroyed by fire which was neither the responsibility of the building owner nor the painter. There are other cases where it will be a question of degree whether the contract is frustrated. The fact that a contractor experiences greater difficulty in carrying out the contract or that it costs it far more than the contractor could reasonably have expected is not sufficient grounds for frustration. Neither will a contract be frustrated by the occurrence of some event which the contract itself contemplated and for which it made provision: *Wates* v. *Greater London Council* (1983). In practice, it is very rare for a contract to be frustrated.

12.1.4 Breach

A breach of contract which is capable of bringing the obligations of both parties to the contract to an end is termed a repudiatory breach. A breach of this nature must strike at the very root of the contract: *Photo Production* v. *Securicor Ltd* (1980). The offending party must clearly demonstrate that it does not intend to accept its obligations under the contract. Such an instance under a building contract could take place where the employer, without any very good reason, prevents all further access onto the site and engages another contractor to complete the work. That would be a very clear repudiation by the employer. Many acts of repudiation are less obvious.

There is no common law right for any party to treat a contract as repudiated simply because the other party is in breach of the obligation to pay. Consistent failures to pay, however, such that a party has lost all confidence of ever being paid may be repudiation in certain circumstances: *D R Bradley (Cable Jointing) Ltd* v. *Jefco Mechanical Services* (1989). More recently it has been held that if a party fails to pay one instalment and intimates that no further monies will be paid until completion, it will amount to repudiation: *C J Elvin Building Services Ltd* v. *Noble* (2003).

A repudiatory breach by one party does not automatically end the contract. The innocent party has the right to affirm the contract and claim damages arising from the breach, or it may accept the breach, bringing its obligations under the contract to an end, and claim damages.

12.2 Termination generally

The contractual provisions for termination do not provide for one or other of the parties to bring the contract to an end, because to do so would mean that all the clauses in the contract (with the exception of the arbitration and adjudication clauses: *Heyman* v. *Darwins* (1942); *Connex South Eastern Ltd* v. *M J Building Services Group plc* (2005)) would fall. The contract expressly provides for termination of the contractor's *employment* under the contract. This puts beyond all doubt that the clauses dealing with consequences of termination continue to apply. 'Termination' has the meaning of 'cessation' or 'conclusion'. It has been introduced into all JCT contracts to replace the word 'determination' which had been used for many years but which is capable of a wider meaning than 'termination'.

The termination clauses in DB are very similar to the equivalent clauses of SBC. The very similarity induces a misplaced familiarity which, in turn, can lead to difficulties.

It is essential to remember that the grounds for termination under this contract would not all amount to repudiatory breaches at common law. It is useful to have a specific contractual machinery for termination, because to rely on common law repudiation can be very uncertain. However, contractual termination must not be thought to end all problems in that respect, because deciding whether the precise grounds have been satisfied can bring its own problems of interpretation of the clauses and of the facts.

In certain instances, where the facts give the innocent party the choice, it may pay it to accept the breach as repudiatory at common law rather than proceed to operate the termination mechanism under the contract. This is because the acceptance of repudiation entitles the party to damages, whereas to terminate under the contract simply entitles the party to whatever remedies the contract stipulates. This may not always be sufficient: *Thomas Feather & Co (Bradford) Ltd* v. *Keighley Corporation* (1953). In rare cases, a party may be able to rely on the contractual termination provisions and acceptance of repudiation in the alternative: *Architectural Installation Services* v. *James Gibbons Windows Ltd* (1989). Termination under the contract and repudiation are usually considered to be mutually exclusive. That is because to terminate under the contract may be considered to be an affirmation of the contract, whereas a party accepting repudiation is saying that the terms of the contract no longer govern either party. It may be possible to overcome the problem in a suitable case by accepting the repudiation first and subsequently operating the termination provisions without prejudice to the repudiation.

The principal termination clause under this contract is clause 8. Termination may also occur under clause 6.10.2.2 and schedule 3, paragraph C.4.4.

Very usefully, clause 8 gathers at the beginning certain provisions which are applicable to all the termination situations. Very importantly, clause 8.2.1 makes clear that the termination must not be carried out unreasonably or vexatiously. 'Vexatiously' suggests an ulterior motive to oppress or annoy: *John Jarvis Ltd* v. *Rockdale Housing Association Ltd* (1986). 'Unreasonably' has been held to be 'taking advantage of the other side in circumstances in which, from a business point of view, it would be totally unfair and almost smacking of sharp practice': *Hill* v. *London Borough of Camden* (1980). Clause 8.2.2 clarifies a sometimes obscure point that termination takes effect when the termination notice is received. The importance of that clause lies in the fact, in regard to other liabilities, that it is sometimes crucial to know to the exact day when termination took place.

Clause 8.2.3 states that all the notices in clause 8 must be in writing and given by actual, special or recorded delivery. In view of the need to carefully calculate time periods from the default notice and not to issue the termination notice prematurely, service of the notice by next day special delivery must be advisable. The clause provides that notices given by special or recorded delivery are deemed to have been received on the second business day after posting. (A 'Business Day' is defined in clause 1.1 as a day which is not a Saturday, Sunday or public holiday.) But this is subject to proof to the contrary. In these days of online tracking of special delivery, it can easily be proved that the notice was actually received on the first business day. In the recent case of *Construction Partnership UK Ltd* v. *Leek Developments* (2006), it was held that, under the JCT Intermediate Form of Contract

(IFC 98) with provision for giving notices almost identical to the provision in DB, a default notice given by fax constituted actual delivery for the purposes of the contract. In a sensible judgment, Judge Gilliland said:

> 'A fax, it seems to me, clearly is in writing; it produces, when it is printed out on the recipient's machine, a document, and that seems to me is clearly a notice in writing. The question is, is that actual delivery? It seems to me, if it has actually been received, it has been delivered. Delivery simply means transmission by an appropriate means so that it is received, and the evidence in this case is that the fax has actually been received. There is no dispute as to that. It may not have been read when received, which is a different matter.'

Clause 8.3.1 provides that clauses 8.2–8.7 are without prejudice to other rights and remedies of the employer and clauses 8.9, 8.10 and 8.12 are without prejudice to other rights and remedies of the contractor. In other words, the fact that the employer and the contractor have been given specific rights of termination under the contract does not mean that they cannot decide to use their respective common law rights to terminate if they so wish. This is important in light of the Court of Appeal decision in *Lockland Builders Ltd* v. *John Kim Rickwood* (1995) which suggested that common law rights and the provisions in the contract could only co-exist if the contractor makes clear that it is not going to be bound by the contract or where the contract, as in this instance, contains specific provision. The somewhat later case of *Strachen & Henshaw Ltd* v. *Stein Industrie (UK) Ltd* (1998), also in the Court of Appeal, took a different view.

Clause 8.3.2 provides, somewhat unnecessarily, that despite the grounds of termination, the parties can agree to reinstate the contractor's employment at any time if they so desire; unnecessarily, because the parties can agree more or less whatever they wish without the need for an express term to that effect.

12.3 Termination by the employer

12.3.1 Grounds

Clauses 8.4 to 8.6, 8.11, 6.10.2.2 and paragraph C.4.4 of schedule 3 set out the terms on which the employer may terminate the contractor's employment. There are fifteen grounds for termination, five of which are based on the contractor's default. These five are as follows.

Whole or substantial suspension of carrying out the Works

It may seem strange that, although the equivalent grounds in SBC make express reference to design of the contractor's designed portion, DB confines itself merely to 'carrying out the Works'. However, reference to the definition of 'Works' in clause 1.1 and then reference to the first recital makes clear that 'Works' encompasses design and construction under this form of contract. In practice, it will be difficult to identify suspension of design work. Construction can be observed progressing (or not progressing) on the site. Design normally takes place in an office somewhere, quite possibly at some distance from the contractor's own office. The employer would usually only know that design work was suspended if specifically informed. In order to qualify under these grounds, the suspension must be complete or substantial. It must also be without reasonable cause.

It could be said that it would be reasonable for the contractor to suspend work if it was waiting for some consent or approval which the employer must give. The contractor will be entitled to suspend for failure to pay under the Housing Grants, Construction and Regeneration Act 1996. This power is exercised under the contract clause 4.11, which expressly excludes it from treatment as a suspension under these grounds. Whether the contractor's suspension was reasonable would depend on the facts in each case. It would certainly be dangerous for the employer to rely on the contractor's suspension as grounds for termination if the contractor could put up a convincing argument, even if suspension for that reason was not strictly in accordance with the contract. To be reasonable in this context, it may not be necessary to be empowered by the contract.

Failure to proceed regularly and diligently

The meaning of 'regularly and diligently' has been considered in section 7.2. These grounds are a breach of the contractor's obligation to comply with clause 2.3. Guidelines were laid down in *West Faulkner Associates* v. *London Borough of Newham* (1995) which help to show whether the contractor is failing in this regard. Needless to say, suspension under clause 4.11 is not a failure to proceed regularly and diligently for the purposes of these grounds.

Failure to remove defective work

Four criteria must be satisfied before these grounds can be invoked:

- The employer must have given a written notice to the contractor; and
- The notice must require the removal of defective work, materials or goods; and
- The contractor must have refused or neglected to comply; and
- As a result the Works must be substantially affected.

These grounds relate to breach of an instruction given by the employer under clause 3.13.1. At first sight, termination appears to be a draconian measure in response, for example, to the contractor's failure to remove defective door handles. It is doubtful whether such a failure would qualify as substantially affecting the Works. Clearly what is indicated is a failure on the part of the contractor which is deliberate ('refuses or neglects') and which seriously affects the Works. Such an instance might occur where the contractor deliberately neglects to remove some defective work which is about to be covered up and possibly may be required to give support to further work, and the subsequent rectification will cause significant delay and expense. Even if the contractor appeared reluctant, the employer could use the powers under clause 3.6 to engage others to carry out the rectification without terminating the contractor's employment.

Failure to comply with the assignment and sub-contracting clauses

The contractor's obligations under clauses 3.3 and 7.1 are to obtain the employer's consent before assignment of the contract or sub-letting. Assignment without consent would be ineffective: *St Martins Property Corporation Ltd and St Martins Property Investments Ltd* v. *Sir Robert McAlpine & Sons Ltd* (1992). In some circumstances, failure to obtain consent to sub-letting will be very serious. This is especially the case under this contract where a contractor attempts to sub-let the design of the Works without consent. It is suggested that termination will be the last resort in any event.

Failure to comply with the CDM Regulations

This is wide in scope and it effectively gives a termination remedy for failure under the relevant contract clauses to comply with the Regulations. These are the clauses grouped under clause 3.18.

12.3.2 Procedure

An employer who intends to terminate the contractor's employment must follow the procedure precisely. If the contractor defaults in one of the five ways noted above, the employer must give the contractor a written notice which must specify the default. It is thought that, strictly, the default is one of the five noted above and not the precise circumstances of the default, which is more correctly the evidence supporting the allegation of default. For example, the employer need only say that the default consists in failure to proceed regularly and diligently. There is no need to say that the contractor had only three men on site for a month and that less than a thousand pounds' worth of work was carried out in that time, etc. It may not be in the employer's interests to give too much detail at this stage. However, it is necessary that the contractor is left in no doubt about the default alleged and the employer must give sufficient detail to identify the incident if there is any danger of confusion: *Wiltshier Construction (South) Ltd* v. *Parkers Developments Ltd* (1997).

As noted earlier, all notices are to be given by actual delivery, special or recorded delivery. The employer should take care to follow the notice procedure precisely.

If the contractor continues the default for 14 days after it receives such notice, the employer may within a further 10 days terminate its employment by serving a notice of termination.

If the contractor stops the default within the stipulated 14-day period or if the employer fails to issue a notice of termination, but subsequently the contractor commits the same default again, perhaps months later, the employer may terminate its employment straightaway or within a reasonable time of the repetition without the need for a further notice. The employer must take great care if attempting to put this provision into effect. The default must be precisely the same default which prompted the employer to serve the original default notice. If that criterion can be met, the employer would be sensible to serve a warning letter to the contractor, before issuing the termination notice. Although this is not an express contractual requirement, indeed the contract appears to state the contrary, the serving of a further letter giving notice of perhaps only 3 days' duration, would serve to counter any possibility that the employer could be said to be acting unreasonably. It is important that the employer states in the letter that it is not a notice of default. That is because a further default notice would be invalid: *Robin Ellis Ltd* v. *Vinexsa International Ltd* (2003).

12.3.3 Insolvency and corruption

These grounds for termination are under the terms of clauses 8.5 and 8.6 respectively.

Insolvency

Clause 8.1 defines insolvency:

- The contractor makes a composition or an arrangement with its creditors.
- The contractor, without declaring insolvency, makes a determination or passes a resolution to wind up.
- The contractor has a winding up or bankruptcy order made against it.
- The contractor under the Insolvency Act 1986 has an administrator or an administrative receiver appointed.
- The contractor is the subject of a similar event or proceeding in another jurisdiction.
- If a partnership, each partner is the subject of an arrangement or other event as above.

What this all amounts to, in simple terms, is that the contractor becomes insolvent. Legal advice is indicated.

The employer may terminate the contractor's employment at any time, but it should be noted that termination is no longer automatic. The contractor must notify the employer in writing immediately any event connected with the clause 8.1 insolvency items takes place. Clause 8.5.3 is very important. It establishes that the date the contractor becomes insolvent is the key date, not the date on which the employer terminates the contractor's employment.

Corruption

Clause 8.6 provides for termination if the contractor is guilty of corrupt practices. The employer should terminate by the issue of a written notification. There is no requirement that the employer should serve a preliminary notice, because it matters not that the contractor has stopped the corruption. The employer has the power to terminate, but is not obliged to do so. Although perhaps academic, it is curious that the employer's power is stated to be to terminate the contractor's employment 'under this or any other contract'. This phrase refers to 'any other contract' between the same two parties. Even with that proviso, it is difficult to see how the parties can bind themselves in law under this contract regarding the substance of other contracts. Indeed, they may not be able to do so. It would require a similar clause in other contracts to entitle the employer to terminate the contractor's employment under those contracts. Although the employer cannot be given power under this contract to terminate the contractor's employment under other contracts, it would be perfectly feasible to insert clauses in this and other contracts to permit the employer to terminate the contractor's employment under any contract if it committed some corrupt act in connection with another contract.

The grounds for termination under this clause are if the contractor, or any person employed by it or acting on its behalf, has committed any offence under the

Prevention of Corruption Acts 1889–1916, or if the employer is a local authority and the contractor has given any fee or reward and receiving it is an offence under sub-section (2) of section 117 of the Local Government Act 1972.

Corruption is a criminal offence and there are severe penalties. The employer is entitled to rescind the contract at common law and to recover any secret commissions. The clause is very strict and it is to be noted that the contractor cannot say in defence under the contract that it did not know what was being done. This clause makes clear that if it was done by the contractor's men, the contractor is liable.

12.3.4 Neutral grounds

Clause 8.11 entitles either party to terminate the contractor's employment if virtually the whole of the Works is suspended due to any of six causes. The period of suspension must be continuous for whatever period the parties stipulate in the contract particulars. If nothing is stated, there is a note which provides that the default period is 2 months, which seems a long time. Users of this contract are advised to insert their own period. It is often difficult to differentiate between suspension of the work and trivial progress. If the site is utterly deserted during the period, there can be no question that work is suspended. If work is progressing to some extent on critical items, it may be wise not to seek to terminate under this clause. However, if some work is progressing, but only on non-critical items and progress is down to a fraction of what it should be, it might well be appropriate to use this clause. Difficulties arise when the suspension stops for a period of a couple of weeks or so then work resumes to near normal and then stops again. That could prejudice the chances of valid termination. The causes are:

- *Force majeure* (see section 8.3).
- Employer's instruction issued under clauses 2.13, 3.9 or 3.10 as a result of the negligence or default of a statutory undertaking carrying out statutory obligations. These instructions deal with divergencies in or between documents, change instructions and postponement.
- Loss or damage due to specified perils. Note that neither party is entitled to terminate under the broader 'all risks' category. The contractor may not terminate under this head if the loss or damage was due to its own negligence or default or to that of any of the contractor's persons.
- Civil commotion or terrorist activity or action of the authorities in dealing with terrorism. Civil commotion is defined as more serious than a riot, but not as serious as a civil war: *Levy* v. *Assicurazioni Generali* (1940).
- The exercise of the UK government of any statutory power directly affecting the Works.
- Delay in receipt of development control permission. These grounds are not to be found in any other JCT contract. Where necessary approvals such as planning permission have not been obtained by the employer before the contract is executed, there is a real risk of the project suffering a long delay before it can commence on site. The contractor must have taken all practicable steps to avoid or reduce the delay, but that probably means little more than that it should have made its application competently and in good time (see also the remarks in section 6.3).

As soon as the stipulated period has expired, all that is required is for the party intending to terminate to serve notice to that effect upon the other by actual, special or recorded delivery that if the suspension is not terminated within 7 days of the date of receipt of the notice, the contractor's employment may be terminated. If the suspension is not ended within the 7 days, the employer or contractor as the case may be may serve a further notice terminating the contractor's employment under the contract.

12.3.5 Terrorism cover and insurance risks

Clause 6.10 deals with the position when terrorism cover ceases to be available. An employer who receives a notice to that effect, either directly or by way of the contractor, may give a written notice to the contractor that the contractor's employment under the contract will terminate. Termination is relatively simple. The notice must state the date of termination, which must be after the date the insurers gave notice (for obvious reasons it could not be prior to this), but before the cessation date of cover stated by the insurers.

Schedule 3, paragraph C.4.4 permits the employer (or the contractor) to terminate the contractor's employment within 28 days of the occurrence of loss or damage to the Works or to any unfixed materials caused by any risks covered by the Joint Names Policy in paragraph C.2. Included is work which is being done to existing structures by way of alteration or extension or both, but the provision does not extend to loss or damage to existing structures or contents. As soon as the damage is discovered, the contractor must give notice in writing to the architect and to the employer, stating the extent, nature and location of the damage. Note that the 28 days begin to run from the occurrence and not from the notification, although in practice if the damage is likely to be sufficiently serious as to form the basis for termination, it will be discovered and notified immediately.

Paragraph C.4.4 states that the termination can only take place 'if it is just and equitable'. This is the crux of the matter. Whether it is just and equitable in any particular instance will depend on all the circumstances. The sort of situation in which termination would be just and equitable would involve such serious damage that it would be uncertain whether work could reasonably recommence. Where a large and expensive building is concerned, undergoing restoration, a large fire could seriously jeopardise any future financial viability. It might be better to demolish and start again. In such a case it will be just and equitable to terminate the contractor's employment. (The contract may also be considered to be frustrated.)

The right of either party to seek adjudication or arbitration on the question of whether it is just and equitable is limited in two ways. The procedures must be invoked within 7 days of receipt of a notice of termination, and the procedure is to decide whether termination will be just and equitable. Whether this attempted restriction will be effective in the case of the right to adjudicate is doubtful.

12.3.6 Consequences

The consequences of termination under clauses 8.4–8.6 are set out in clauses 8.5.3, 8.7 and 8.8.

If due to insolvency, then from the date of the insolvency:

- Clauses 8.7.4 (preparing an account), 8.7.5 (payment) and 8.8 (employer's decision not to complete the Works) apply just as if a termination notice had been given.
- Provisions of the contract requiring any further payment or release of retention cease to apply. The precise boundaries of this provision have become open to doubt since *Melville Dundas Ltd (in receivership)* v. *George Wimpey UK Ltd* (2005). This was a Scottish case concerned with the equivalent of WCD 98. The clauses there were broadly similar to the DB clauses so far as putting a stop to payment is concerned. In the *Melville* case, the court held that the Housing Grants, Construction and Regeneration Act 1996 took precedence over the contract provisions and that the employer could not refuse to pay an outstanding application for payment unless the correct withholding notices had been served.
- The contractor's obligation to complete the Works is suspended (note that it does not cease). However, once the employer has served a notice of termination, the suspension will effectively be a cessation. The key point is that the contractor is not in breach of contract if it stops work on insolvency.
- The employer may take reasonable measures to protect the site, the Works and site materials and the contractor must not hinder the process. It is common for sub-contractors and suppliers who have not been paid to try and recover whatever materials are on the site after the insolvency of the contractor. The employer must take legal advice, but in general it can be said that goods may not be removed if they are fixed to the building. A supplier who entered the site to recover materials would be trespassing, but once the materials are removed, it may be difficult for the employer to gain any kind of redress if the supplier can show that the contractor has not paid for the materials.

Where the termination is under clauses 8.4, 8.5 or 8.6, clause 8.7 allows the employer to engage others to complete the Works, which must of course include design, and make good defects. They may enter and take possession of the site and use all the temporary buildings and plant and all materials (subject to any retention of title clause) and purchase any new materials for that purpose. The contractor is not entitled to remove its temporary buildings etc. from the site until the employer has expressly and in writing required it to do so. Although the contract no longer so states, the contractor must be given a reasonable time in which to remove its buildings and plant, etc. If it does not comply with the employer's request, the employer is probably entitled to remove them without any liability for damage, to sell them and hold the proceeds less costs for the contractor.

Clause 8.7.2.2 requires the contractor to give the employer two copies of all the design documents which have been prepared at the date of termination. It is arguable whether the contractor has any obligation to provide documents which are only partially prepared and which may, at that stage, include errors or discrepancies.

The position of sub-contractors and suppliers is briefly considered under clause 8.7.2.3. If the employer so requires within 14 days of the date of termination, the contractor must assign the benefit of any agreement for the supply of materials or the carrying out of any work to the employer. The assignment is said to be as far as the contractor can lawfully be required to assign. A footnote reminds the user that in the case of the contractor's insolvency, this clause may not be effective. No doubt the sub-contractors and suppliers will normally be pleased at the prospect of assignment if the employer agrees to pay them what they are owed. This may be difficult to achieve and the employer must beware of paying twice for the same thing. However, there appears to be no good reason why a sub-contractor should not be able to charge the employer and receive a premium for finishing its work, if it is specialist work or if circumstances dictate that the same sub-contractor must be used to complete. The employer may take account of such payments in the final account.

Clause 8.7.3 provides that any provision of the contract requiring further payment or release of retention ceases to apply. This deals with termination which is not due to insolvency, but the comments regarding cessation of further payments on insolvency also apply here. Clauses 8.7.4 and 8.7.5 deal with payment. After making good of defects by the contractor employed to complete the Works, an account must be made by the employer. The account must show expenses properly incurred by the employer together with loss or damage caused as a result of the termination, the amount already paid to the contractor and the amount which would have been payable if the Works had been completed in accordance with the contract. The result may be a debt owing to the contractor or to the employer.

If the employer decides not to continue with the Works, clause 8.8 provides that written notification must be send to the contractor within 6 months from the termination. This clause responds to the situation which arose in *Tern Construction Group (in administrative receivership)* v. *RBS Garages* (1993), where the judge had to imply a term. The employer must send the contractor a statement of account within a reasonable time of the notice or, if there is no notice, upon the expiry of the 6-month period. The statement must set out the total value of work properly executed, any other amounts due to the contractor and the amount of expense properly incurred by the employer together with any loss or damage. After taking into account amounts previously paid, the result may be a debt owing to the employer or to the contractor.

There is no express provision requiring the contractor to surrender possession of the site on termination although clause 8.7.1 comes near, but it is considered that the courts would now grant an injunction if the contractor attempted to stand fast. This view is supported by two decisions of Commonwealth courts: *Kong Wah Housing* v. *Desplan Construction* (1991) and *Chermar Productions* v. *Prestest* (1991). These decisions are in sharp contrast to the only English case on the point: *London Borough of Hounslow* v. *Twickenham Garden Developments Ltd* (1970).

12.3.7 Consequences of termination under clauses 8.11, 6.10.2.2 and paragraph C.4.4 of schedule 3

The consequence of termination under these clauses is dealt with in section 12.4.6.

12.4 Termination by the contractor

12.4.1 Grounds

Clauses 8.9, 8.10, 8.11 and paragraph C.4.4 of schedule 3 set out the 12 grounds on which the contractor may terminate its employment under the contract.

The main grounds for termination are divided into two categories: default and suspension.

Default

There are three grounds in the first category, which is to be found in clause 8.9.1:

Employer's failure to pay: The employer must have failed to pay an interim amount properly due. In order for the amount to be properly due it must be contained in an application for payment made under clause 4.9 and it must comply with the provisions of clause 4 (see Chapter 10). It should be noted that it is not sufficient for the contractor to make application for some approximated sum and then to terminate if it is not paid in full. Unlike a traditional contract, the contractor is at the helm so far as calculating payment is concerned. In these circumstances, if termination is to be effective, the contractor must scrupulously carry out its duties. The valuation must be carefully calculated and the appropriate deductions made. If the employer makes any deduction under clause 4.10.4, the contractor must seriously consider whether such deduction is justified before launching into the termination procedure. These are probably the most common grounds for termination and they are certainly valuable, but if the contractor is in error in terminating, it could be held to be in repudiatory breach of contract by its actions although much will depend on the extent to which the contractor has honestly relied on the contract provision, even though it may do so mistakenly: *Woodar Investment Development Ltd* v. *Wimpey Construction UK Ltd* (1980).

Where the contractor desires to terminate on these grounds, the employer must have failed to pay within 14 days of the due date. Reference to clause 4.10.1 makes it clear that the due date is the date of receipt of the contractor's application for payment. In the light of this it makes sense for the contractor to send all applications for payment by special or recorded delivery so that it is possible to obtain proof of receipt if necessary.

Employer fails to comply with the assignment clause: Presumably non-compliance would be an attempt to assign without permission, which we have already seen would be ineffective.

Employer fails to comply with the CDM Regulations: The comments are the same as under employer termination.

Suspension

The second category, which is to be found in clause 8.9.2, contains just one ground for termination, but that ground is extremely broad. The pre-condition is that, after the date of possession but before practical completion, the carrying out of the whole or nearly the whole of the Works has been suspended for the period named in the contract particulars by any impediment, prevention or default of the employer or any of the employer's persons. This is a very similar ground to the grounds for extension of time and loss and/or expense under clauses 2.26.5 and 4.20.5 respectively, and it could scarcely be wider in scope. It is stated in a note to the contract particulars that the period will be 2 months if no period is stated. The contract refers to them as 'specified suspension events'. The suspension must be continuous.

There are two important points the contractor must watch before terminating:

- It must be certain that the suspension has continued for the prescribed period; and
- The suspension must be continuous, i.e. there must be no breaks during which work resumed for a period, no matter how brief.

12.4.2 Procedure

If the contractor intends to terminate, it must serve notice on the employer specifying the default or specified suspension event and requiring an end within 14 days. The contractor must take care not to give notice too early.

If the employer does not end the default by the end of the last day of the 14-day period, the contractor may serve a notice on the employer by actual, special or recorded delivery. The notice should state that notice of termination under clause 8.9.3 will be served if the default or suspension is not ended within 14 days from receipt. The notice will expire on the 15th day, after which the contractor may serve notice of termination by actual, special or recorded delivery. The contractor has 10 days in which to act. If it fails to act within the 10 days or if the default or suspension is ended within the 14-day period, the contractor is entitled under clause 8.9.4 to serve notice of termination if at any time the default is repeated or the suspension event is repeated (even for a short period). No warning notice is prescribed in this instance and the contractor may serve the termination notice immediately or within a reasonable time of such repetition, but it is wise for the contractor to issue a warning letter (but not a further default notice) in any event. This point has been considered above in relation to the employer's right to terminate for a repeated default. The repetition must be precisely the same as the default or suspension which led to the original default notice.

12.4.3 Insolvency

Clause 8.10 provides for termination for the insolvency reasons set out in clause 8.1 and noted above. Once the employer becomes insolvent, the contractor may terminate its employment by sending a written notice to the employer. An employer, who makes any proposal or gives notice of a meeting or an appointment in relation to any of the insolvency events in clause 8.1, has an obligation to write to the contractor immediately. It is important to note that the contractor's obligation to carry out and complete the design and construction of the Works is suspended as soon as any of the insolvency events in clause 8.1 occur, even if the contractor has not served a termination notice at that point.

12.4.4 Neutral grounds

This is clause 8.11. The grounds for termination under this clause have already been covered in section 12.3.4. It should be noted that if the contractor wishes to terminate, there is a proviso in clause 8.11.2 that it is not entitled to give notice if the loss or damage due to specified perils is caused by the contractor's own negligence or default or of contractor's persons. This merely puts into words what would be implied – the contractor cannot gain an advantage through its own default.

12.4.5 Insurance risks

The grounds for termination and the procedure to be adopted where there is a schedule 3, paragraph C.4.4 situation are the same whether the employer or the contractor terminates (see section 12.3.5).

12.4.6 Consequences

Following termination under clauses 6.10.2.2 (employer's right only), 8.9 to 8.11, and paragraph C.4.4 of schedule 3, the rights and duties of the parties are set out in clause 8.12. In essence, the situation is straightforward, as follows:

- Clause 8.12.1 provides that following termination under the above clauses, clause 8.12 applies. The clause goes on to say that other provisions of the contract which require further payment or retention release cease to apply. The comments in section 12.3.6 regarding the withholding of payment also apply here. In referring to the consequences after termination under clause 6.10.2.2, clause 6.10.3 refers to clause 8.12 but excludes the operation of clause 8.12.1. Having done that, clause 6.10.3 proceeds to repeat the substance of clause 8.12.1. It is not at all clear why it was thought necessary to do this and the result appears to be an unnecessary duplication of words.
- The contractor must remove all its temporary buildings, plant, tools, equipment and materials from site. It must act as quickly as is reasonable in the circumstances and, although not expressly stated, it must obviously take whatever precautions are necessary to prevent death, injury or damage. The contractor should give all sub-contractors facilities to remove their equipment etc. Unlike the position under WCD 98, this contract does not expressly prohibit the contractor or sub-contractors from removing goods or materials which have been properly ordered for the Works and for which the contractor has paid or is legally liable to pay. However, reference to clauses 8.12.2.1, 8.12.3.4 and 8.12.5 makes clear that the cost of such goods and materials is to be included in the account and, on payment, they become the property of the employer. Unlike the position where the employer terminates as a result of the contractor's default, this provision makes clear that the contractor is neither entitled nor obliged to leave its property on the site until it feels like moving it or until it has another site to receive it or until the employer instructs.
- Clause 8.12.2.2 refers to the contractor's duty to provide the employer with two copies of the as-built drawings referred to in clause 2.37 prepared at the date of termination. It should be noted that this requirement is not the same as clause 8.7.2.2 which provides that the contractor must provide copies of all the design documents prepared at the date of termination. This requirement is more obscure, because as-built drawings are not the same as the drawings prepared for

construction and it is not usual for as-built drawings to be completed until the Works are built. Therefore, it may be that if termination takes place, as it would, during the progress of the Works, as-built drawings may not have been prepared and there may be nothing for the contractor to provide under this clause.

- Clause 8.12.3 stipulates that an account must be prepared. If the termination has been as a result of employer's default or insolvency under clauses 8.9 or 8.10, the contractor must prepare the account. The clause says that the contractor must do it as soon as it is reasonably practical. Obviously, the contractor will do it immediately. It will probably arrive on the employer's desk the day after termination. If termination is by either party under clause 8.11 or paragraph C.4.4 of schedule 3 or by the employer under clause 6.10.2.2, it is for the employer to state whether the employer or the contractor is to prepare the account. If the employer opts to prepare it, the contractor must provide all the information necessary for the account within 2 months of termination. The employer must, at the latest, prepare the account within 3 months of receipt of the documents. The 3 months is clearly the absolute maximum period allowed by the contract. The key phrase is that the employer must prepare the account 'with reasonable dispatch'. In most instances, the contractor's idea of documents to enable the employer to prepare an account will amount to the contractor's final account with all supporting information. In order to avoid procrastination on the part of the employer, it is suggested that the contractor provides this as soon as possible after the termination. Then, whichever option is chosen (even belatedly) by the employer will have been satisfied by the contractor and the onus will be on the employer either to prepare an employer's account within 3 months of receipt of the contractor's account or to accept the contractor's account.
- The contents of the account are stipulated in clause 8.12.3. The payments are to consist of the value of all construction work which is properly executed and all design work carried out, amounts due for direct loss and/or expense under clauses 3.17 and 4.19 including amounts ascertained after the date of termination, the contractor's reasonable costs incurred in removal from site, amounts in respect of materials for which the contractor has paid or is legally liable to pay, and direct loss and/or damage which the contractor has incurred as a result of the termination. In an appropriate case, such loss and/or expense could include the whole of the profit which the contractor would have made had it been allowed to complete the contract: *Wraight Ltd* v. *P H & T Holdings* (1968). This would be subject to the contractor demonstrating that, on the balance of probabilities, it would have made a profit. Claims for loss of profit are not sustainable in the abstract, particularly when all the evidence points to the contractor having made a loss and continuing to do so: *McAlpine Humberoak* v. *McDermott International Inc* (1992). It should be noted that Clause 8.12.4 restricts the contractor's entitlement to such direct loss and/or damage to the consequences of termination under clauses 8.9 or 8.10 or under clause 8.11.1.3 if the loss or damage due to specified perils resulted from the employer's or the employer's persons' negligence or default.
- Within 28 days of submission of the account by one of the parties to the other in accordance with the contract, and after taking into account any sums previously paid to the contractor under this contract, the amount properly due must be paid by either the contractor or the employer to the other, but specifically without the deduction of any retention.

12.4.7 Consequences of termination under paragraph C.4.4 of schedule 3

Where either party terminates under this paragraph, all the provisions of clauses 8.12.2 to 8.12.5 apply except clause 8.12.3.5 (see section 12.4.6). Clause 8.12.3.5 deals with direct loss and/or damage caused by the termination and its omission in this instance is a consequence of the fact that the termination is not based on fault.

It cannot be emphasised too strongly that termination should be treated as a last resort. Even if all the pitfalls in the process are successfully negotiated, the result is likely to be expensive for both parties.

Chapter 13
Dispute Resolution

13.1 *General*

Four methods of dispute resolution are referred to in the contract, although only three need to be considered here.

13.1.1 Mediation

Little need be said about the first method: mediation. It is dealt with by clause 9.1. The clause is very brief. It simply states that, by agreement, the parties may choose to resolve any dispute or difference arising under the contract through the medium of mediation. A footnote refers users to the Guide. It is not clear why this clause has been included in the contract at all. It was not in the WCD 98 edition except in the form of a footnote. The key to the redundant nature of this clause is in the phrase 'The Parties may by agreement . . .'. The parties, of course, may do virtually anything by agreement. They can agree to set aside the whole contract and sign a different one if they are both of one mind on the matter. One assumes that this clause was inserted purely to remind the less sophisticated users of the form that mediation is a possibility. If that is the explanation, it is not immediately obvious why the draughtsman of the contract did not also refer to the possibility of conciliation or negotiation. There is little point in including as terms of a contract anything which is to be agreed. The whole point of a written contract is that it is evidence of what the parties have already agreed. To have a clause which effectively states: 'we may agree to do something else' is a complete waste of space. It is to be hoped that future editions of the contract consign this clause either back to a footnote or, better still, to the Guide where it belongs.

13.1.2 Adjudication

The Housing Grants, Construction and Regeneration Act (commonly called the Construction Act) (the Act) was enacted in 1996. (In Northern Ireland Part II of the Act is virtually identical to the Construction Contracts (Northern Ireland) Order 1997.) Section 108 of the Act expressly introduces a contractual system of adjudication to construction contracts. Excluded from the operation of the Act are contracts relating to work on dwellings occupied or intended to be occupied by one of the parties to the contract. DB, in common with other standard forms, incorporates the requirements of the Act. Therefore, all construction Works carried out under this form are subject to adjudication even if they comprise work to a dwelling house. Briefly, section 108 provides that:

- A party to a construction contract has the right to refer a dispute under the contract to adjudication.
- Under the contract:
  - — A party can give notice of intention to refer to adjudication at any time
  - — An adjudicator should be appointed and the dispute referred within 7 days of the notice of intention
  - — The adjudicator must make a decision in 28 days or whatever period the parties agree
  - — The period for decision can be extended by 14 days if the referring party agrees
  - — The adjudicator must act impartially
  - — The adjudicator may use initiative in finding facts or law
  - — The adjudicator's decision is binding until the dispute is settled by legal proceedings, arbitration or agreement
  - — The adjudicator is not liable for anything done or omitted in carrying out the functions unless in bad faith.
- If the contract does not comply with the Act, the Scheme for Construction Contracts (England and Wales) Regulations 1998 (the Scheme) will apply.

The right to refer to adjudication 'at any time' means that adjudication can be commenced even if legal proceedings (and presumably arbitration) are in progress about the same dispute: *Herschel Engineering Ltd* v. *Breen Properties Ltd* (2000). It has also been held that adjudication can be sought even if repudiation of the contract has taken place. A dispute may be referred to adjudication and the adjudicator may give a decision even after the expiry of the contractual limitation period: *Connex South Eastern Ltd* v. *M J Building Services Group plc* (2005). Of course, in such a case, the referring party runs the risk that the respondent will use the limitation period defence, in which case the claim will normally fail.

Paragraph C.4.4.1 of schedule 3 states that either party may 'within 7 days of receiving [a termination notice] (*but not thereafter*) invoke the dispute resolution procedures that apply . . .' [emphasis added] and it appears to contravene the requirements of the Act that a party has the right to refer to adjudication 'at any time'. The clause attempts to restrict the period during which a referral may take place. It would almost certainly be ineffective against the adjudication clause.

Adjudication is dealt with in article 7 and clause 9.2. It might be termed a 'temporarily binding solution'. In the vast majority of cases, it seems that the parties accept the adjudicator's decision and do not take the matter further. Even where there are challenges through the courts against the enforcement of an adjudicator's decision, the challenge is concerned with matters such as the adjudicator's jurisdiction or whether the adjudicator complied with the requirements of natural justice, not whether the adjudicator's decision was correct. Although it may seem odd, the courts cannot interfere with the adjudicator's decision, no matter how obviously wrong, provided that the adjudicator has answered the questions posed by the referring party. In *Bouygues United Kingdom Ltd* v. *Dahl-Jensen United Kingdom Ltd* (2000), the Court of Appeal said:

> 'The first question raised by this appeal is whether the adjudicator's determination in the present case is binding on the parties . . . The answer to that question turns on whether the adjudicator confined himself to a determination of the issues that were put before him by the parties. If he did so, then the parties are bound by his determination, notwithstanding that he may have fallen into error. As Knox J put it in *Nikko Hotels (UK) Ltd* v. *MEPC plc* [1991] 2 EGLR 103 . . . if the adjudicator has answered the right question in the wrong way, his decision will be binding. If he has answered the wrong question, his decision will be a nullity.'

Quite so. This view was re-stated in the Scottish courts: *Gillies Ramsey Diamond* v. *PJW Enterprises Ltd* (2003).

The parties must comply with the adjudicator's decision following which, if they are not satisfied, either party may instigate proceedings through the stipulated system of obtaining a final decision. It is important to remember that, in doing so, the parties are not appealing against the decision of the adjudicator and the arbitrator or court will ignore the adjudicator's previous decision in arriving at an award or judgment respectively: *City Inn Ltd* v. *Shepherd Construction Ltd* (2000).

13.1.3 Arbitration

Only two of the dispute resolution methods produce a final and binding decision. The choice is between arbitration and legal proceedings. It should be noted that, unlike the position under previous contracts up to and including WCD 98, legal proceedings will apply unless the contract particulars are completed to show that arbitration is to be the dispute resolution procedure. The decision to default to legal proceedings rather than arbitration seems to have been taken after significant lobbying by solicitors, who traditionally feel at home with court proceedings and, with a few exceptions, less happy with arbitration. The advantages of arbitration are often said to be:

- *Speed*: This depends to a great extent on the arbitrator; a good arbitrator should dispose of most cases in months, not years.
- *Privacy*: Only the parties and the arbitrator are privy to the details of the dispute and the award. Of course, sometimes the threat of publicity is a useful tactic against an unscrupulous party.
- *The parties decide*: The parties can decide timescales, procedure and location of any hearing; that is, if they can agree anything at this stage.
- *Expense*: Theoretically, arbitration should be more expensive than litigation because the parties (usually the losing party) have to pay for the arbitrator and the hire of a room, but in practice the speed and technical expertise of the arbitrator usually keep costs down.
- *Technical expertise of the arbitrator*: The fact that the arbitrator understands construction should shorten the time schedule and possibly avoid the need for expert witnesses if the parties agree.
- *Appeal*: The award is final because the courts are loath to consider any appeal: *The Council of the City of Plymouth* v. *D R Jones (Yeovil) Ltd* (2005).

Possible disadvantages are:

- In theory, it is more expensive because the parties (usually the losing party) pay the cost of the arbitrator and the hire of a room for the hearing. That is not a problem for the successful party, of course.
- If the arbitrator is ineffective, the process may be slow and expensive and fail to produce a good result.
- The arbitrator may not be an expert on the law, which may be a major part of the dispute. The result may be a defective award. The answer is for the parties at least to agree on the arbitrator.
- Parties who are in dispute often find it difficult to agree about anything. Therefore, the arbitrator may be appointed by the appointing body and the procedure, the timing and the location of the hearing room may be decided by the arbitrator, with the result that neither party is satisfied.

Arbitration is probably still the most satisfactory procedure for the resolution of construction disputes and employers would be advised to complete the contract particulars accordingly. If the parties have agreed that the method of binding dispute resolution will be arbitration, a party who attempts to use legal proceedings instead will fail in a costly way if the other party relies on section 9 of the Arbitration Act 1996: *Ahmad Al-Naimi* v. *Islamic Press Agency Incorporated* (2000). Section 9 requires the court to grant a stay (postponement) of legal proceedings until the arbitration is concluded unless the arbitration is null, void, inoperable or incapable of being performed. The court has no discretion about the matter and the successful party will claim its costs. The result is not only that the party intent on legal proceedings will have to revert to arbitration, but it will have to pay the other party's legal costs in opposing the legal proceedings.

Arbitration is dealt with by article 8 and clauses 9.3 to 9.8.

13.1.4 Legal proceedings

The advantages of legal proceedings are said to be:

- The judge is an expert on the law. On the other hand, many judges are overturned on appeal.
- Many of the judges in the Technology and Construction Courts have a sound understanding of construction matters.
- The Civil Procedure Rules require judges to manage their caseloads and encourage pre-action settlement through use of the Pre-Action Protocol. This may end in adjudication rather than legal proceedings.
- Cases can reach trial quickly. People who have been through the legal system are not always convinced about this.
- The claimant can join several defendants into the proceedings to allow interlocking matters and defendants to be decided.
- Costs of judge and courtroom are minimal.
- A dissatisfied party can appeal to a higher court.

The disadvantages of legal proceedings are said to be:

- Even specialist judges know relatively little about the details of construction work.
- Parties cannot choose the judge, who may not be very good or at any rate unsuitable for the case.
- Costs will be added because expert witnesses or a court-appointed expert witness will be needed to assist the judge to understand relatively simple points.
- Cases often take years to resolve.
- Lengthy timescale and complex processes may result in high costs.
- Appeals may result in an unacceptable level of costs.

Many contracts are filled in quite casually and the likelihood is that, as a result, legal proceedings will become the norm for any parties wishing to have a final binding decision. Legal proceedings are dealt with by article 9.

13.2 Adjudication

13.2.1 Under the contract

The contract provides, in article 7, that if any dispute or difference arises under the contract, either party may refer it to adjudication in accordance with clause 9.2. The parties are the employer and the contractor. The employer's agent is not a party to the contract even though the name of the agent will be written into article 3. Usually, the contractor will initiate adjudication, although there is nothing to stop the employer from doing so. For example, the employer may seek adjudication if disagreeing with the contractor's application for payment.

The employer's agent is not a party to the contract and, therefore, cannot be the respondent to an adjudication. The employer's agent can act as a witness, but the agent has no duty to run an adjudication on behalf of the employer. Acting in an adjudication on behalf of either the employer or the contractor usually calls for some degree of skill and experience which most construction professionals, acting in the normal course of their professions, will not readily acquire. The problem is that what was originally intended to be a simple and readily accessible system has been rendered highly complex by the multitude of court decisions about every aspect of the procedure. Although the court performs an invaluable service by interpreting the statute, the Scheme and various contract clauses, it is inevitable that specialist skills are then required to understand how all these decisions fit together. Where the dispute is other than very straightforward or where one party has retained the services of a legal representative, the other party is well advised to do likewise. That is certainly the case if there is any question of the existence of a dispute or the jurisdiction of the adjudicator.

Only disputes arising 'under' the contract may be referred. This can be contrasted with arbitration where the disputes are described in article 8 as 'of any kind whatsoever arising out of or in connection with this Contract'. This phrase has been considered (*Ashville Investments Ltd* v. *Elmer Contractors Ltd* (1987)), and held to be very broad in meaning. Thus, for example, an adjudicator has no power to consider formal settlement agreements about various matters made by the parties in connection with the contract of which the adjudication clause forms part (*Shepherd Construction* v. *Mecright Ltd* (2000)), but a written variation to written contract terms is part of the underlying contract and any dispute about whether it is enforceable is one arising under the contract (*Westminster Building Co Ltd* v. *Beckingham* (2004)).

It is not necessary to refer a dispute to adjudication before seeking arbitration or legal proceedings, as the case may be, although many members of the construction industry believe that to be the case. Use of the word 'may' makes clear that adjudication is optional. Nevertheless, it is rapidly replacing arbitration as the standard dispute resolution process, albeit that it is rather rough and ready. Unfortunately, it is often used for complex disputes involving large amounts of money for which it is not suited. Adjudications involving £1 million or more with time extended by agreement to 2 or 3 months are a travesty of what the Act intended and it is hoped that adjudicators will take a strong line with such referrals. The author has frequently received referrals consisting of many full lever arch files. It is not clear how the referring party in such instances believes the adjudicator is to read and absorb the information within the time available. The adjudicator has wide powers to set the procedure and can direct the parties to reduce the volume of paper submitted if that is appropriate.

It will be noted that clause 9.2 is much shorter than the comparable clause in WCD 98. That is because the detailed procedure set out in WCD 98 has been abandoned in favour of the procedure in the Scheme. This is a very sensible approach because the Scheme is a comprehensive set of rules especially drafted to comply with the Act. Currently, there is a proliferation of procedures, none of them offering any serious advantages to the Scheme.

Use of the Scheme is made subject to certain provisos:

- The adjudicator and nominating body are to be those stated in the contract particulars.
- If the dispute concerns whether an instruction issued under clause 3.13.3 (opening up or testing following work found not to be in accordance with the contract) is reasonable, the adjudicator, if practicable, must be someone with appropriate expertise and experience. If not, the adjudicator must appoint an independent expert with appropriate expertise and experience to give advice and to report in writing whether the instruction is reasonable in all the circumstances.

Why clause 3.13.3 should have been singled out is unclear. The adjudicator must always have relevant expertise and experience or seek expert assistance.

13.2.2 The Scheme: the notice of adjudication

In paragraph 1, the Scheme provides that any party to a construction contract may give to all the other parties a written notice of an intention to refer a dispute to adjudication. The notice must describe the dispute and the parties involved, and must give details of the time and location, the redress sought and the names and addresses of the parties to the contract. If appropriate, the address to which notices should be sent, as specified by the parties, must be included. This last is very important. It is more common than might be thought for a party to a contract to change its address from the one written in the contract. The referring party has to provide information so that the adjudicator can correspondent immediately with the parties.

It has been known for a referring party to send the notice, and subsequently the referral, to the address in the contract knowing that the other party has moved, in the hope of putting the other party at a disadvantage. Most adjudicators are alive to this kind of nonsense and a referring party acting in that way will probably cause the adjudicator to question the validity of its case. Clause 1.6 of the contract, which deals with the service of documents, makes clear that, if not agreed, documents must be sent to the last known business address or registered office.

The notice of adjudication is the trigger for the adjudication process and it is also one of the most important documents. Great care must be taken in its preparation because the dispute which the adjudicator is entitled to consider is the dispute identified in the notice of adjudication: *McAlpine PPS Pipeline Systems Joint Venture* v. *Transco plc* (2004); *Carillion Construction Ltd* v. *Devonport Royal Dockyard Ltd* (2005) CA. The dispute cannot be broadened later by the referring party, although it can be elaborated and more detail provided: *Ken Griffin and John Tomlinson* v. *Midas Homes Ltd* (2000). In that case, the court set out the purposes of the notice as:

> '. . . first, to inform the other party of what the dispute is; secondly, to inform those who may be responsible for making the appointment of an adjudicator, so that the correct adjudicator can be selected; and finally, of course, to define the dispute of which the party is informed, to specify precisely the redress sought, and the party exercising the statutory right and the party against whom a decision may be made so that the adjudicator knows the ambit of his jurisdiction.'

For example, if the notice of adjudication states the dispute as being the amount due in an application for payment and if the redress sought is simply the adjudicator to decide the amount due, the adjudicator will have no power to order payment of that amount although, doubtless, that would be what the referring party wishes: *F W Cook Ltd* v. *Shimizu (UK) Ltd* (2000) Again, it must be emphasised that the adjudicator can only answer the question posed in the notice of adjudication. The adjudicator is not empowered to answer the question which should have been asked, however frustrating it may be, and an adjudicator who tried to do that would be acting in excess of jurisdiction. The decision would be a nullity. Sometimes an adjudicator will see the omission in the notice and will ask the parties if they wish to give jurisdiction for the adjudicator to consider the omitted question; paragraph 20 allows the adjudicator to take into account any other matters which the parties agree should be within the adjudication's scope. Unless both parties agree to confer jurisdiction, the adjudicator must simply answer the question posed even though it means that a second adjudication is inevitable.

The adjudicator is expressly empowered to take into account matters which the adjudicator considers are necessarily connected with the dispute. To take a simple example, it is probably essential for an adjudicator to decide the extent of extension of time allowable, even if not asked, before deciding about the amount of liquidated damages properly recoverable. The express empowerment merely puts into words what would be the legal position in any event: *Karl Construction (Scotland) Ltd* v. *Sweeney Civil Engineering (Scotland) Ltd* (2002); *Sindall Ltd* v. *Solland* (2001).

13.2.3 The Scheme: appointing the adjudicator

Paragraphs 2–6 set out the procedure for selecting an adjudicator. The process is relatively complex, which reflects the general nature of the Scheme.

The appointment of the adjudicator under the Scheme cannot take place until after the notice has been served: *IDE Contracting Ltd* v. *R G Carter Cambridge Ltd* (2004). Paragraph 2 contains the framework and it is made subject to the overriding point that the parties are entitled to agree the name of an adjudicator after the notice of adjudication has been served. If the parties can so agree, they have the best chance of an adjudicator who has the confidence of both parties. Sadly, parties in dispute find it difficult to agree on anything at all. If there is a person named in the contract, that person must first be asked to act as adjudicator. There are difficulties with having a named person: the person may be away or ill or even dead when called upon to act; the person's expertise may be unsuitable for the particular dispute, or pressure of work may force that person to decline.

The contract makes provision for an adjudicator to be named in the contract particulars. If there is no named person or if that person will not or cannot act, the referring party must ask the nominating body indicated in the contract particulars to nominate an adjudicator. There are problems with this approach also. Not all adjudicators are of equal capability. Indeed, some of them are far from satisfactory. Some have a tenuous grasp of the law while many others wrongly believe that the adjudicator's job is to make decisions according to their own gut-feeling notions of right and wrong without reference to law. If the referring party asks for a nomination, both parties are stuck with the result unless they agree to revoke the appointment. However, for the reason already stated, such an agreement is unlikely. The nominating body is to be stated in the contract particulars. The bodies listed are:

- Royal Institute of British Architects
- Royal Institution of Chartered Surveyors
- Construction Confederation
- National Specialist Contractors Council
- Chartered Institute of Arbitrators.

It is important that four of the bodies should be deleted. If there is no adjudicator named and no body is selected, the referring party may choose any one of the bodies on the list to make the nomination. If there is no list of appointing bodies, perhaps because all of them have inadvertently been deleted, the referring party is free to choose any nominating body to make the appointment.

A nominating body is fairly broadly defined in paragraph 2(3) of the Scheme as a body which holds itself out publicly as a body which will select an adjudicator on request. The body may not be what is referred to as a 'natural person', i.e. a human being, nor one of the parties. Paragraph 6 states that if an adjudicator is named in the contract, but for some reason cannot act or does not respond, the referring party has three options. The first is to ask any other person specified in the contract to act, the second is to ask the adjudicator nominating body in the contract to nominate, and the third is to ask any other nominating body to nominate. It will readily be seen that this procedure is simply a clarification of existing options.

From receiving the request, the nominated adjudicator has 2 days in which to accept. The adjudicator must be a person and not a body corporate. Therefore, a firm of quantity surveyors cannot be nominated although one of the directors or partners can be nominated. Paragraph 5 provides that the nominating body has 5 days from receipt of the request to communicate the nomination to the referring party. Invariably, a nominating body will also notify the respondent, but surprisingly the Scheme does not expressly require such notification.

In the event that the nominating body fails to nominate within 5 days, the parties may either agree on the name of an adjudicator or the referring party may request another nominating body to nominate. In either case the adjudicator has 2 days to respond, as before.

One of the parties may object to the identity of the adjudicator. Paragraph 10 makes clear that the objection of either party to the adjudicator will not invalidate the appointment, nor any decision reached by the adjudicator. It is useful to have this spelled out. There is a misconception among the uninitiated that a party has only to register an objection to the adjudicator in order to bring the process to an end or at least suspend it. Nothing could be further from the truth. It is entirely a matter for each party the extent to which it wishes to participate in the adjudication. If a party objects, it should make quite clear that any participation is without prejudice to that position and to the party's right to refer the objection to the courts in due course. It may be catastrophic for a party knowing of grounds for objection to continue the adjudication without further comment. In such cases, the party may well be deemed to have accepted the adjudicator: *R Durtnell & Sons Ltd* v. *Kaduna Ltd* (2003).

Elaborate provision is made in paragraphs 9 and 11 if the adjudicator resigns or the parties revoke the appointment. The provisions are sensible. The adjudicator may resign at any time on giving notice in writing to the parties. It should be noted that the notice need not be reasonable and, therefore, immediate resignation is possible. The referring party may serve a new notice of adjudication and seek the appointment of a new adjudicator as noted above. If the new adjudicator so requests and if reasonably practicable, the parties must make all the documents available which have been previously submitted.

If an adjudicator finds that the dispute is essentially the same as a dispute which has already been the subject of an adjudication decision, that adjudicator must resign: *Holt Insulation Ltd* v. *Colt International Ltd* (2001). It is not always easy to decide whether a dispute has been referred before in precisely that form. An adjudicator may be entitled to consider an extension of time dispute already adjudicated if the second referral identified additional grounds for delay: *Quietfield Ltd* v. *Vascroft Construction Ltd* (2006) CA. Because this is a jurisdictional question, the adjudicator does not have jurisdiction to decide it but has to decide it nevertheless as a matter of practicality, i.e. whether to proceed or to stop. The adjudicator is entitled to determine a reasonable amount due by way of fees and expenses and how it is to be apportioned. The parties are jointly and severally liable for any outstanding sum. Although, oddly, not given as a reason for resignation, the significant variation of a dispute from what was referred in the referral notice, so that the adjudicator is not competent to decide it, is a trigger for entitlement to payment. At first sight, this criterion is not easy to understand. The only sensible interpretation is that it appears to refer to a significant difference between the dispute as set out in the notice of adjudication and what the referring party then includes in the referral notice.

Paragraph 11 provides for revocation of the appointment by the parties, but it does not seem to be a common occurrence. When it occurs, the adjudicator is entitled to determine a reasonable amount of fees and expenses and the apportionment. The parties, as before, are jointly and severally liable for any balance. Parties will find it difficult to challenge the amount of fees determined by an adjudicator unless the adjudicator can be shown to have acted in bad faith. It is not sufficient to show that a court would have arrived at a different sum: *Stubbs Rich Architects* v. *W H Tolley & Son* (2001).

13.2.4 The Scheme: procedure

Paragraph 3 stipulates that a request for the appointment of an adjudicator must include a copy of the notice of adjudication. As the court stated in the *Ken Griffin* case, this is to assist those making the nomination and the prospective adjudicator so that a suitable person is nominated.

There is little time available because, in compliance with the Act, paragraph 7 stipulates that the referring party must submit the dispute in writing to the adjudicator, with copies to each party to the dispute, no later than 7 days after the notice of adjudication. This submission is known as the 'referral notice'. Looking at the procedure for appointment of the adjudicator, it is clear that the timetable is tight. The referral notice, which effectively is the referring party's claim, must be accompanied by relevant parts of the contract and whatever other evidence the referring party relies on in support of the claim. If the referral is served later than the 7th day, it will be invalid and the adjudicator will have no jurisdiction: *Hart Investments Ltd* v. *Fidler and Another* (2006). In an earlier case, the court considered that the adjudicator had limited discretion to accept a referral submitted 1 or 2 days after the 7 days had expired: *Floyd Slaski Partnership* v. *London Borough of Lambeth* (2001).

It is surprising that the Scheme does not even state that the respondent may reply to the referral notice, let alone set out a timescale as was the case in WCD 98. No doubt reliance is placed on the fact that, in order to comply with the rules of natural justice, the adjudicator is obliged to allow a reasonable period for the reply. The provisions in WCD 98 used to allow 7 days for the reply. Respondents always believe this to be totally inadequate to reply to what may be a referral notice and a considerable amount of evidence. It is a matter for the adjudicator to decide, but in view of the restricted overall period for the decision it seems that 14 days is the very most which any respondent can expect. Usually, the adjudicator will allow less than this. Much will depend on the complexity of the dispute and the volume of relevant paper, bearing in mind that the respondent will almost certainly have a detailed knowledge of the project.

Paragraph 19 provides that the adjudicator must reach a decision 28 days after the date of the referral notice. Note that this is from the date of the notice, not from the date the notice is received by the adjudicator. The period may be extended by 14 days if the referring party consents or, if both parties agree, for any longer period. Where a substantial sum of money is being claimed, an adjudicator should be alive to the fact that, if the claim is found to be valid, it means that a contractor has been kept out of its money unlawfully for an extended period. The award of interest, although better than nothing, rarely compensates for the default. An adjudicator should try to arrive at a decision before the 28 days expires and to resist asking for an extension of the period unless unavoidable. If the adjudicator does not comply with the timetable in reaching the decision, either party may serve a new notice of adjudication and request a new adjudicator to act. The new adjudicator can request copies of all documents given to the former adjudicator. Paragraph 19(3) requires the adjudicator to deliver a copy of the decision to the parties as soon as possible after the decision has been reached.

Doubts have been raised about the position if the adjudicator does not reach a decision in time or if the decision reached in time is not delivered as soon as possible. There are two conflicting Scottish decisions and two English decisions dealing with these questions. In brief, *Ritchie Brothers PWC Ltd* v. *David Philp (Commercials) Ltd* (2005) decided that the adjudicator's jurisdiction to make a decision ceased on the expiry of the time limit if not already extended. *St Andrews Bay Development Ltd* v. *HBG Management Ltd* (2003) decided that if the adjudicator reached a decision within the relevant timescale, a 2-day delay in delivering the decision to the parties was not sufficient to render the decision a nullity. *Simons Construction Ltd* v. *Aardvark Developments Ltd* (2004) decided that the decision of an adjudicator is binding, even if given after the expiry of the relevant timescale, provided that neither party has served a new notice of adjudication before the decision has been reached. *Barnes & Elliott Ltd* v. *Taylor Woodrow Holdings Ltd* (2004) decided that if the adjudicator reached the decision within the relevant timescale, a short delay in communicating it to the parties was within the tolerance and commercial practice that should be afforded to the Act and the contract.

The Scottish decisions are not, of course, binding in English courts, but they may be persuasive if there is no English decision on the point. Since there are English decisions, the position appears to be as set out in those decisions: a late or late-communicated decision is valid provided neither party has taken steps to bring the adjudication to an end, such as serving a fresh notice of adjudication after the expiry of the relevant period.

If one of the parties fails to comply with the adjudicator's decision, the other party may seek enforcement of the decision through the courts. The courts will normally enforce the decision unless there is a jurisdictional or procedural problem. In enforcement proceedings, the court is not being asked to comment on the adjudicator's decision or reasoning, although a court will quite often do so, thus obscuring the *ratio* of the judgment. Where a court is asked to enforce an adjudicator's decision, the important part of the judgment is simply the reasons why the judge decided to enforce or not. Any comments the judge may make on the adjudicator's decision itself will be *obiter;* at best persuasive, but certainly not of binding force.

13.2.5 The Scheme: adjudicator's powers and duties

Only one dispute may be referred to adjudication at the same time under the Scheme: *Fastrack Contractors Ltd* v. *Morrison Construction Ltd and Another* (2000). However, paragraph 8 permits the adjudicator to adjudicate at the same time on more than one dispute under the same contract, provided that all parties consent. The adjudicator may deal with related disputes on several contracts even if not all the parties are parties to all the disputes, provided they all consent. Moreover, the parties may agree to extend the period for decision on all or some of the disputes.

It is clear that multiple dispute procedures bring their own complications for which the Scheme, wisely, does not try to legislate. For example, it is not clear whether multiple disputes, certainly under different contracts, must be adjudicated on under one big adjudication. The wording of paragraph 8 seems to leave the position open. Where there are different contracts and the parties vary from one contract to another, it will be a matter of discussion and agreement whether the adjudicator should conduct separate adjudications albeit at the same time. It is quite conceivable that parties to different contracts may not wish others to know some of the details of the dispute. Of course, if the disputes were being dealt with through the courts, not only the parties themselves but any member of the public would be able to sit in and listen to those details. Another problem with multiple adjudications is that an adjudicator may have been appointed for one or more of the adjudications. If so and that adjudicator ceases to act under this paragraph, the adjudicator concerned is entitled to render a fee account to the relevant parties who will be jointly and severally liable for its discharge if no apportionment is made or if one party fails to pay some or all of its share.

The adjudicator's duties are to act impartially in accordance with the relevant contract terms, to reach a decision 'in accordance with the applicable law in relation to the contract' and to avoid unnecessary expense. Sadly, some adjudicators seem to be unaware of their obligations to apply the law to their decisions, and decisions are made on the basis of the adjudicator's idea of fairness, moral rights or justice. The author has had the misfortune to see some appalling adjudicator's decisions. In one instance the adjudicator decided that the executed contract was unsuitable for the Works and drafted the decision as though a different contract had been used. Needless to say, the decision was declared a nullity by the court. On another occasion, an adjudicator declared himself a 'watchdog for fairness'. That is a misguided view of the adjudicator's role, which has been stated to be 'primarily to decide the facts and apply the law (in the case of an adjudicator, the law of the contract)': *Glencot Development & Design Company Ltd* v. *Ben Barrett & Son (Contractors) Ltd* (2001). Fortunately, there are also some very good adjudicators with a clear understanding of their roles.

The Scheme gives the adjudicator some very broad and some very precise powers:

- To take the initiative in ascertaining the facts and the law.
- To decide the procedure in the adjudication.
- To request any party to supply documents and statements.
- To decide the language of the adjudication and order translations.
- To meet and question the parties.
- To make site visits, subject to any third party consents.
- To carry out any tests, subject to any third party consents.
- To obtain any representations or submissions.
- To appoint experts or legal advisors, subject to giving prior notice.
- To decide the timetable, deadlines and limits to length of documents or oral statements.
- To issue directions about the conduct of the adjudication.

Paragraph 14 requires parties to comply with the adjudicator's directions. If a party does not comply, the adjudicator has significant powers under paragraph 15:

- To continue the adjudication notwithstanding the failure.
- To draw whatever inferences the adjudicator believes are justified in the circumstances.
- To make a decision on the basis of the information provided and to attach whatever weight to evidence submitted late the adjudicator thinks fit.

Parties who are unused to adjudication often wonder whether it is necessary to be represented or whether they can handle the adjudication themselves. The simple answer is that there is nothing to prevent a party from acting personally in the adjudication, but it is usually wise to seek representation from an experienced professional. Paragraph 16 of the Scheme deals with representation. A party may have assistance or representation as deemed appropriate, but there is a proviso that oral evidence or representation may not be given by more than one person unless the adjudicator decides otherwise.

Sometimes, although it is becoming less common, a contract may provide that a decision or certificate is final and conclusive. Unless that is the case, the adjudicator is given power to open up, revise and review any decision or certificate given by a person named in the contract. It is worth noting that to be exempt from revision by the adjudicator, the decision or certificate must be stated to be both final and conclusive. A contract which simply states that a certificate is conclusive is open to review. On that basis, the final account and final statement are exempt because they are called 'final' and, unless they are disputed within the stipulated timescale, they are also conclusive. Obviously they are reviewable if the reference is made before the expiry of 28 days from the date of issue, in accordance with clause 1.9.2.

The adjudicator is also given power to order any party to the dispute to make a payment, to state its due date and the final date for payment and to decide the rates of interest, the periods for which it must be paid and whether it must be simple or compound interest. In deciding what, if any, interest must be paid, the adjudicator must have regard to any relevant contractual term. To 'have regard' to a contractual term is a rather loose phrase which probably means little more than to give attention to it. It falls short of the need to actually comply with it: *R* v. *Greater Birmingham Appeal Tribunal* ex parte *Simper* (1973).

The adjudicator must consider relevant information submitted by the parties and if the adjudicator believes that other information or case law should be taken into account, it must be provided to the parties and they must have the opportunity to comment: *Balfour Beatty Construction Ltd* v. *London Borough of Lambeth* (2002). Paragraph 18 puts a prohibition on the disclosure of information, noted by the supplier as confidential, to third parties by any party to an adjudication or the adjudicator unless the disclosure is necessary for the adjudication.

13.2.6 The Scheme: the adjudicator's decision

The construction industry as a whole does not adequately understand that the whole adjudication process from service of notice to compliance with the adjudicator's decision is intended to be conducted with great dispatch. Paragraph 23 empowers the adjudicator to order the parties to comply peremptorily with the whole or any part of the decision. In the absence of any directions about the time to comply, paragraph 26 makes clear that compliance must be immediate on delivery of the decision to the parties. One of the problems is that enforcement through the courts is still not usually available at short notice and all legal process is expensive to both parties. The Scheme repeats the requirement of the Act that the decision will be binding and must be complied with until the dispute is finally determined by arbitration, legal proceedings or agreement. Occasionally, the losing party will try to avoid payment by attempting to set off money allegedly owed under the contract or elsewhere by the successful party against the amount decided by the adjudicator. This approach is unlikely to be successful in light of the Court of Appeal decision in *Ferson Contractors Ltd* v. *Levolux AT Ltd* (2003) which held that where it appeared that the obligation to pay the amount decided by the adjudicator might be thwarted by reference to the contract, the obligation to pay took precedence. Exceptions to this principle will be rare: *R J Knapman Ltd* v. *Richards and Others* (2006).

In contrast to the position under WCD 98, paragraph 22 provides that if either party so requests, the adjudicator must give reasons for the decision. Some adjudicators purport not to give reasons, but simply indications or limited reasons. The court in *Joinery Plus Ltd (in administration)* v. *Laing Ltd* (2003) had some useful things to say about reasons:

> 'The statement by the adjudicator that he was only giving reasons for a limited purpose or was only giving limited reasons has little if any practical effect. If an adjudicator gives any reasons, they are to be read with the decision and may be used as a means of construing and understanding the decision and the reasons for that decision. There is no halfway house between giving reasons and publishing a silent or non-speaking decision without any reasons. There is no way in which reasons may be given for a limited purpose and which are only capable of being used for that purpose.'

The court went on to say that comments about the decision given by the adjudicator after delivering the decision were irrelevant except to the extent that the adjudicator was entitled to correct basic mistakes in the decision, if invited to do so: *Bloor Construction (UK) Ltd* v. *Bowmer & Kirkland (London) Ltd* (2000). However, the reasons given may be brief:

> 'If an adjudicator is requested to give reasons pursuant to paragraph 22 of the Scheme, in my view a brief statement of those reasons will suffice. The reasons should be sufficient to show that the adjudicator has dealt with the issues remitted to him and what his conclusions are on those issues. It will only be in extreme circumstances . . . that the court will decline to enforce an otherwise valid adjudicator's decision because of the inadequacy of the reasons given. The complainant would need to show that the reasons were absent or unintelligible and that, as a result, he had suffered substantial prejudice.' (*Carillion Construction Ltd* v. *Devonport Royal Dockyard Ltd* (2005))

The adjudicator is entitled to reasonable fees, which the adjudicator may determine. The parties are jointly and severally liable for payment if the adjudicator makes no apportionment or if there is an outstanding balance.

Paragraph 26 states that the adjudicator will not be liable for anything done or omitted in carrying out the functions of an adjudicator unless the act or omission is in bad faith. Similar protection is also given to any employee or agent of the adjudicator. It is perhaps worth noting that, as an incorporated term of the contract, this paragraph is not binding on persons who are not parties to the contract.

13.2.7 The Scheme: costs

Nothing in the Scheme allows the adjudicator to award the parties costs. This is in harmony with the philosophy of the Act, which does not encourage the parties to incur large amounts of costs in pursuing claims. In arbitration and litigation, by contrast, where costs are normally awarded against the losing party, the dispute can deteriorate into a fight about costs rather than the point at issue. That has much to do with the huge costs which can be incurred by each side.

Despite this, whether or not, in a particular instance, an adjudicator can award costs has caused problems. *John Cothliff Ltd* v. *Allen Build (North West) Ltd* (1999), which decided that the Scheme gave the adjudicator power to award costs, was considered in *Northern Developments (Cumbria) Ltd* v. *J & J Nichol* (2000). There the court did not agree with the earlier case and in a concise judgment held that there was no provision in the Scheme which gave the adjudicator such power. However, the adjudicator could be given power to award costs, either expressly by the parties or by implied agreement. In that case, both parties had professional representation. Both parties asked the adjudicator to award costs and neither party made any submissions that the adjudicator had no power to award costs. As the judge said: 'It would have been open to either party to say to the Adjudicator, I have only asked for costs in case you decide that you have jurisdiction to award them but I submit that you have no jurisdiction to make such an award.' Where the parties have agreed that the adjudicator is to decide and apportion the legal costs of the parties, the adjudicator may, depending on all the circumstances, still retain the power to do so even after the referring party has discontinued an adjudication so as to remove the need for the adjudicator to make a decision about it: *John Roberts Architects Ltd* v. *Parkcare Homes Ltd* (No.2) Ltd (2006).

13.3 Arbitration

13.3.1 General

Until the House of Lords' decision in *Beaufort Developments (NI) Ltd* v. *Gilbert Ash NI Ltd* (1998), it was considered that a court did not have the same power as the arbitrator, conferred by the contract, to open up and revise certificates and decisions. This was a result of the judgment in *Northern Regional Health Authority* v. *Derek Crouch Construction Co Ltd* (1984). Undoubtedly, arbitration was the dispute resolution procedure of choice, rather than legal proceedings, in many contracts for that reason alone. That was because both employers and contractors were afraid that if matters were left to litigation, disputes about certification (which the employer as well as the contractor might wish to challenge) could not properly be resolved.

That approach was changed by the *Beaufort* case which held that a court has the same powers as an arbitrator to open up, review and revise certificates, opinions and decisions of the architect. Indeed, the court has the power as a right, whereas the power must be conferred on the arbitrator by the parties. It has been remarked at the beginning of this chapter that legal proceedings are now the default procedure in DB, rather than arbitration. Now that it is clear that the judge and the arbitrator have similar powers to open up certificates and the like, the parties may still wish to retain arbitration for some of the other reasons set out earlier.

DB arbitration procedures are brief. Arbitration can take place on any matter at any time. Arbitrators appointed under a JCT arbitration agreement are given extremely wide express powers. Their jurisdiction is to decide any dispute or difference of any kind whatsoever arising under the contract or connected with it (article 8). The scope could scarcely be broader (*Ashville Investments Ltd* v. *Elmer Contractors Ltd* (1987)) and by clause 9.5 the arbitrator's powers extend to:

- Rectification of the contract to reflect the true agreement between the parties.
- Directing the taking of measurements or the undertaking of such valuations as the arbitrator thinks desirable to determine the respective rights.
- Ascertaining and making an award of any sum which should have been included in a certificate.
- Opening up, reviewing and revising any certificate, opinion, decision, requirement or notice issued, given or made and determining all matters in dispute as if no such certificate, opinion, decision, requirement or notice had been given.

The JCT 2005 edition of the Construction Industry Model Arbitration Rules (CIMAR) current at the contractual base date, is stated to govern the proceedings (clause 9.3). The provisions of the Arbitration Act 1996 are expressly stated by clause 9.8 to apply to any arbitration under this agreement. That is to be the case no matter where the arbitration is conducted. Therefore, even if the project and the arbitration take place in a foreign jurisdiction, the UK Act will apply provided that the parties contracted on DB, and clause 1.11 referring to the law of England is not amended.

The following matters are specifically excluded from arbitration:

- Disputes about value added tax.
- Disputes under the Construction Industry Scheme, where legislation provides some other method of resolving the dispute.
- The enforcement of any decision of an adjudicator.

The employer and contractor agree, by clause 9.7 in accordance with sections 45(2)(a) and 69(2)(a) of the Act, that either party may by proper notice to the other and to the arbitrator apply to the courts to determine any question of law arising in the course of the reference, and appeal to the courts on any question of law arising out of an award. When the clause was originally introduced, it was viewed with some doubt on the basis that the courts might not accept it as satisfying the requirements prior to such an appeal. However, clauses like this have been held to be effective: *Vascroft (Contractors) Ltd* v. *Seeboard plc* (1996).

Arbitration, like litigation, is almost always costly in terms of both money and time. No matter how powerful and convincing the case may be, there is no guarantee of success. Even the successful party will often look back at the cost, the time spent and the mental stress involved and conclude that it was not worth the effort. There are always some people who will threaten arbitration over trivial matters in an attempt to gain an advantage. Indeed, it is a recognised, although possibly ineffective, form of negotiation to suddenly abandon talks and serve an arbitration notice. Unfortunately, even with the recent review of dispute resolution procedures and introduction of the adjudication process, that approach will not disappear. It will not always be possible to avoid arbitration and, therefore, employers and contractors must ensure that they properly appreciate how the process operates. Only then can they recognise the possible consequences of embarking on formal arbitration proceedings.

It is quite commonly thought that the arbitration process is a fairly informal get-together to enable the parties and the arbitrator to have a chat about the dispute before the arbitrator decides, in a consensual kind of a way, who should be successful. That is much more a description of a conciliation meeting. In fact, the majority of arbitrations are conducted quite formally, like private legal proceedings – which is what they are. The arbitration begins by inviting the parties to the 'preliminary meeting', but that does not mean a friendly discussion. It is a formal meeting to establish all the important criteria which need to be decided before the arbitration can proceed. The arbitrator will usually work from an agenda. Sometimes parties attempt to gain an advantage by springing a surprise request on the arbitrator at that meeting. Experienced arbitrators have no difficulty in dealing with such requests, but there is a limit to the degree to which the arbitrator can ensure that one party is not disadvantaged by such tactics. A party should not go to a preliminary meeting without taking a fully briefed legal advisor experienced in arbitration.

The employer and contractor are free to agree who should be appointed as, or should appoint, the arbitrator and they have freedom to agree important matters such as the form and timetable of the proceedings. This raises the possibility of a quicker procedure than would otherwise be the case in litigation and even matters such as the venue for any future hearing might be arranged to suit the convenience of the parties and their witnesses.

If oral evidence and cross-examination are to be carried out, it is usually done at a hearing. Hearings, which are the private equivalent of a trial, are conducted in private, not in an open court. Parties are free to choose whether to represent themselves or whether to be represented and by whom. They need not, in the traditional courtroom way, be represented by solicitor and counsel. However, it is not usually advisable for the parties to represent themselves because difficult legal points can arise in apparently the simplest of arbitrations. Some arbitrations are won by clever tactics, particularly where there is a weak arbitrator. Therefore, experienced help is essential.

13.3.2 Procedure

Arbitrations begun under DB and subject to the law of England must be conducted subject to and in accordance with the JCT 2005 edition of CIMAR, current at the base date of the contract. If any amendments have been issued by JCT since that date, the parties may jointly agree to give written notice to the arbitrator to conduct the reference according to the amended rules (clause 9.3).

CIMAR rule 2.1 and section 14(4) of the Arbitration Act 1996 provide that arbitral proceedings are begun in respect of a dispute when one party serves on the other a written notice of arbitration 'identifying the dispute' and requiring agreement to the appointment of an arbitrator.

CIMAR is a comprehensive body of rules, generally of admirable clarity. It consists of a set of rules with two appendices, the first defining terms and the second helpfully reproducing sections of the Arbitration Act 1996 which are relevant but not already included in the rules. These are followed by the JCT Supplementary and Advisory Procedures, the first part of which is mandatory and must be read with the rules. The greater part is advisory only, but appears to be well worth adopting. At the back of the document is a set of notes prepared by the Society of Construction Arbitrators dated 1 February 1998, updated January 2002. The whole document, at the time of writing, is available on www.jctcontracts.com. As might be expected, JCT/CIMAR is very detailed and repays careful study. Among other things, it offers the parties a choice of three broad categories of procedure by which the proceedings will be conducted:

- Short hearing procedure
- Documents only procedure
- Full procedure.

They must be considered separately.

Short hearing procedure (rule 7)

This procedure is not very common although it can be useful in certain instances. It limits the time available to the parties within which to orally address the matters in dispute before the arbitrator. Although the time can be extended by mutual consent, in the absence of that agreement no more than one day will be allowed during which both parties must have a reasonable opportunity to be heard. Before the hearing, either by simultaneous exchange or by consecutive submissions, each party will provide to the arbitrator and to each other a written statement of their claim, defence and counterclaim (if any). Each statement must be accompanied by all relevant documents and any witness statements on which it is proposed to rely. The JCT procedures usefully insert some timescales for certain of the steps.

If it is appropriate to do so, either before or after the short hearing, the arbitrator may inspect the subject matter of the dispute if desired. This procedure is particularly suited to issues which can fairly easily be decided by such an inspection of work, materials, plant and/or equipment or the like. The arbitrator must decide the issues and make an award within a month after concluding hearing the parties.

It is possible to present expert evidence. However, it is costly and often unnecessary. That is particularly the case if the arbitrator has been chosen specifically on the grounds of specialist knowledge and expertise. Parties can sometimes agree to allow the arbitrator to use that specialist expertise when reaching the decision and so the use of independent expert evidence under the short hearing procedure is all but actively discouraged under rule 7.5, which precludes any party calling such expert evidence from recovering the costs of doing so, except where the arbitrator determines that such evidence was 'necessary for coming to his decision'.

This procedure with a hearing is ideally suited to many common disputes which are relatively simple and it provides for a quick award with minimum delay and associated cost.

Documents only procedure (rule 8)

This will only be a viable option if all the evidence is contained in the form of documents. Nevertheless, when the criteria are satisfied, it can offer real economies of time and cost. It is only suited to disputes which are capable of being dealt with in the absence of oral evidence and where the sums in issue are relatively modest and do not warrant the time and associated additional expense of a hearing. Parties, in accordance with a timetable devised by the arbitrator, will serve on each other and on the arbitrator a written statement of case, which as a minimum will include:

- An account of the relevant facts and opinions upon which reliance is placed.
- A statement of the precise relief or remedy sought.

If either party is relying on evidence of witnesses of fact, the relevant witness statements (called 'proofs'), signed by the witnesses concerned, will be included with the statement of case. If the opinions of an expert or experts are required, they must similarly be given in writing and signed. There is a right of reply and if there is a counterclaim, the other party may reply to it.

Despite the procedure being called 'documents only', the arbitrator may set aside up to a day during which to question the parties and/or their witnesses if it is considered desirable. The arbitrator must make a decision within a month or so of final exchanges and questioning, but there is provision for the arbitrator to notify the parties that more time for the decision will be required. The JCT procedures again set out a useful timetable.

Full procedure (rule 9)

If neither of the other options is considered satisfactory, CIMAR makes provision for the parties to conduct their respective cases in a manner similar to conventional High Court proceedings, offering the opportunity to hear and cross-examine factual and expert witnesses.

This is the most complex procedure and the JCT procedure which sets out a detailed timetable for various activities within the procedure is of real assistance to the parties and to the arbitrator. It is intended that the rules will accommodate the whole range of disputes which might arise. Therefore, they offer a sensible framework for conducting the proceedings. They may be modified as appropriate so that they can be used effectively and efficiently for the particular dispute under consideration.

The unamended rules lay down that parties will exchange formal statements. In difficult or complex cases, the statements will comprise the claim, defence and counterclaim (if any), reply to defence, defence to counterclaim and reply to defence to counterclaim. Each submission must be sufficiently detailed to enable the other party to answer each allegation made. As a minimum, the statement must set out the facts and matters of opinion which are to be established by evidence. It may include statements concerning any relevant points of law, evidence or reference to the evidence that it is proposed will be presented, if this will assist in defining the issues and a clear statement of the relief or remedies sought.

The date of commencement of the arbitration is important for two particular reasons. The first is relevant in terms of the Limitation Act 1980. If the notice is served before the expiry of the limitation period, it prevents the respondent using the limitation defence. The second reason is that notice of arbitration often results in a counterclaim from the respondent. It seems doubtful, strictly, whether any counterclaim might be brought within the jurisdiction of the arbitration which has already commenced, without a formal process being executed.

It is common for respondents simply to raise their counterclaims formally at or about the time of serving their defences to the primary claim. That may not be possible in the face of an attack by a claimant wishing to frustrate the respondents' attempts to automatically bring that counterclaim into the proceedings. It should be noted that CIMAR rule 3.2 allows any party to an arbitration to give notice in respect of any other dispute. Provided it is done before the arbitrator is appointed, the disputes are to be consolidated. Rule 3.3 allows either party to serve notice of any other dispute after the arbitrator has been appointed, but consolidation is not then automatic. Rule 3.6 of CIMAR makes clear that arbitral proceedings in respect of any other dispute are begun 'when the notice of arbitration for that other dispute is served'. Although a claimant's insistence that the respondent serve a fresh notice to cover a counterclaim may only be a temporary inconvenience to the respondent, there are serious practical issues.

If there is a doubt concerning whether a counterclaim has properly been brought within the original arbitration, it may affect the existence and the extent to which either party has protection from liability for costs. This is especially the case if previous 'without prejudice' offers of settlement have been made. If it is long after the initial arbitration has been commenced that the respondent realises that a fresh notice is necessary to pursue the counterclaim, the consequences could be serious not only in terms of costs, but also in regard to the limitation period.

It is one of the important tasks of the arbitrator to give detailed directions concerning everything necessary for the proper conduct of the arbitration. Often, those directions will include orders regarding the time within which either party may request further and better details of the other party's case and the reply to any such request. Directions may also be given requiring the disclosure of any documents or other relevant material which is or has been in each party's possession. Probably, the parties will be required to exchange written statements setting out any evidence that may be relied on from witnesses of fact in advance of the hearing. There will also be directions given regarding any expert witnesses, the length of the hearing or hearings and the time available for each party to present it case.

13.3.3 The appointment of an arbitrator

It is the option of either party to begin arbitration proceedings. As a first step, one party must write to the other requesting them to concur in the appointment of an arbitrator (clause 9.4.1). Whoever does so, proceedings are formally commenced when the written notice is served. Rule 2.1 of CIMAR sets out the procedure, stating that the notice must identify the dispute and require agreement to the appointment of an arbitrator. It is good practice for the party seeking arbitration to insert the names of three prospective arbitrators. This saves time and often both parties can agree on one of the names. If the respondent maintains that none of the names is acceptable, it is usual for that party to volunteer a new set of names. The arbitrator must have no relationship to either of the parties, nor should they have connections with any matter associated with the dispute.

It is important for the parties to make a sincere effort to agree on a suitable candidate rather than having one appointed whose skills and experience may be entirely unknown. Depending on the nature and amount of money at stake in the dispute, the parties may be unwilling to agree anything. In that case, clause 9.4.1 of the contract and rule 2.3 of CIMAR provide that if the parties cannot agree on a suitable appointment within 14 days of a notice to concur or any agreed extension to that period, either party can apply to a third party to appoint an arbitrator. There is a list of appointors in the contract particulars against clause 9.4.1. All but one should be deleted leaving the agreed appointor as either: the President or a Vice-President of the Royal Institute of British Architects, the Royal Institution of Chartered Surveyors, or the Chartered Institute of Arbitrators. If no single body has been chosen, the default provision is the President or a Vice-President of the Royal Institute of British Architects. Of course it is always open to the parties to the contract to insert the name of a different appointor of their choice at the time the contract is executed. It will be necessary to complete special forms and to pay the relevant fee. Although the system of appointing an arbitrator varies, the aim is the same. The object is to appoint a person of integrity who is independent, having no existing relationships with either party or their professional advisors and who is impartial. It should go without saying that the arbitrator should have the necessary and appropriate technical and legal expertise. Claimants who have a dispute to refer and respondents receiving a notice to concur should waste no time in taking proper expert advice on how best to proceed.

If the arbitrator's appointment is made by agreement, it will not take effect until the appointed person has confirmed willingness to act, irrespective of whether terms have been agreed. If the appointment is the result of an application to the appointing body, it becomes effective, whether or not terms have been agreed, when the appointment is made by the relevant body (CIMAR rule 2.5). There is no fixed scale of charges for arbitrator's services, and fees ought to depend on their experience, expertise and often on the complexity of the dispute. Arbitrators commonly charge between £1,000 and £2,000 a day. They usually require an initial deposit from the parties and, if there is to be a hearing, there will be a cancellation charge graded in accordance with the proximity of the cancellation to the start of the hearing. The argument in support of this is that the arbitrator will have put 1 or 2 weeks on one side for the hearing, during which time no other work has been booked. A cancellation means that it is difficult for the arbitrator to secure work at short notice to fill the void. In cases where the cancellation fee is substantial, due to proximity to the hearing date, it might be sensible to ask the arbitrator to account to the parties for any fee-earning activities during the hearing period, with which to discount the cancellation fee.

When appointed, the arbitrator will consider which of the procedures summarised above appears to be most appropriate as a forum for the parties to put their cases. The arbitrator must choose the format that will best avoid undue cost and delay, and that is often a most difficult balancing act. Therefore, parties must within 14 days after acceptance of the appointment is notified to the parties, provide the arbitrator with an outline of their disputes and of the sums in issue, along with an indication of which procedure they consider best suited to them.

After due consideration of all parties' views and unless a meeting is considered unnecessary, the arbitrator must, within 21 days of the date of acceptance, arrange a meeting (the preliminary meeting) which the parties or their representatives will attend to agree (if possible), or receive the arbitrator's decision on, everything necessary to enable the arbitration to proceed. It is obviously preferable for the parties to agree which procedure is to apply. If they cannot agree, the documents only procedure will apply unless the arbitrator, after having considered all representations, decides that the full procedure will apply.

The parties are always free to conduct their own cases, but if disputes have reached the stage of formal proceedings it is usually better to engage experienced professionals to act for them.

13.3.4 Powers of the arbitrator

Under the 1996 Arbitration Act the arbitrator's powers have been significantly broadened. For example, an arbitrator may:

- Order which documents or classes of documents should be disclosed between and produced by the parties: section 34(2)(d).
- Order whether the strict rules of evidence shall apply: section 34(2)(f).
- Decide the extent to which the arbitrator should take the initiative in ascertaining the facts and the law: section 34(2)(g).
- Take legal or technical assistance or advice: section 37.
- Order security for costs: section 38.
- Give directions in relation to any property owned by or in the possession of any party to the proceedings which is the subject of the proceedings: section 38.
- Make more than one award at different times on different aspects of the matters to be determined: section 47(1).
- Award interest: sections 49(1)–49(6).
- Make an award on costs of the arbitration between the parties: sections 61(1) and 61(2).
- Direct that the recoverable cost of the arbitration, or any part of the arbitral proceedings, is to be limited to a specified amount: sections 65(1) and 65(2).

13.3.5 Third party procedure

An advantage of litigation over arbitration is that claimants can take action against several defendants at the same time and any defendant can seek to join in another party who may have liability. This facility is not readily available in arbitration, which usually takes place only between the parties to the contract. In DB, the draughtsman has wisely left the possibility of a multi-party arbitration to be covered by the CIMAR rules. However, the courts have shown that they are not adverse to such multi-party proceedings where otherwise there would be substantial costs generated: *City & General (Holborn) Ltd* v. *AYH plc* (2006).

Rules 2.6 and 2.7 provide that where there are two or more related sets of proceedings on the same topic, but under different arbitration agreements, anyone who is charged with appointing an arbitrator must consider whether the same arbitrator should be appointed for both. In the absence of relevant grounds to do otherwise, the same arbitrator is to be appointed. If different appointors are involved they must consult one another. If one arbitrator is already appointed, that arbitrator must be considered for appointment to the other arbitrations.

This situation commonly occurs when there is an arbitration under the main contract and also between the contractor and a sub-contractor about the same issue, perhaps one of valuation or extension of time. It is also possible that there are two contracts between the same two parties and an issue arises in both which is essentially the same point. Usually, the same arbitrator ought to be appointed for that situation.

13.4 Legal proceedings

This option is dealt with by article 9 which simply provides that the English courts will have jurisdiction over any dispute or difference arising out of or in connection with the contract. Parties wishing to adopt this procedure will delete the arbitration option (article 8). It should be remembered that the default position has changed under this contract. If neither option is deleted, legal proceedings are the default position.

Table of Cases

The following abbreviations are used:

AC	Law Reports, Appeal Cases
ALJR	Australian Law Journal Reports
ALR	Australian Law Reports
All ER	All England Law Reports
App Cas	Law Reports, Appeal Cases
BCL	Building and Construction Law (Australia)
BLISS	Building Law Information Subscriber Service
BLM	Building Law Monthly
BLR	Building Law Reports
B & S	Best & Smith's Bench Reports
C & P	Carrington & Payne Reports
Ch	Law Reports, Chancery Division
CILL	Construction Industry Law Letter
CLD	Construction Law Digest
Con LR	Construction Law Reports
Const LJ	Construction Law Journal
DLR	Dominion Law Reports
EG	Estates Gazette
EGLR	Estates Gazette Law Reports
EGCS	Estates Gazette Case Summaries
ER	English Reports
EWCA Civ	England and Wales Court of Appeal (Civil Division)
EWHC	England and Wales High Court
Ex	Law Reports, Exchequer Division
JP	Justice of the Peace
KB	Law Reports, King's Bench Division
LGR	Local Government Reports
LT	Law Times Reports
Lloyds Rep	Lloyds Law Reports
NZLR	New Zealand Law Reports
QB	Law Reports, Queen's Bench Division
SLT	Scottish Law Times
TLR	The Times Law Reports
UKPC	United Kingdom Privy Council
WLR	Weekly Law Reports

Clause Index

Subject Index